CHE

June 19

TILL

COME HOME

Philip Walling started out farming in Cumbria, then trained as a barrister and practised for twenty-five years, before turning to writing. From the law he brings learning and rigour, while his roots in the land give him a passion for and deep understanding of the landscape and people of rural England – a combination which lends a unique perspective to his work. He now lives in Northumberland. His first book, *Counting Sheep*, was a *Sunday Times* bestseller.

'A well-researched, uncynical and thought-provoking book. It is also an elegy to a dying industry.' Nigel Farndale, *The Times*

'A vital, thorough and accessible history that everyone who cares about the past or the future should read.' Rosamund Young, *Sunday Times* bestselling author of *The Secret Life of Cows*

'Entertaining... A colourful and informative romp through the history of native breeds, brought to life through delightful anecdotes.' *Country Life*

'Engaging... Lively... Fascinating' *Spectator*

'A hymn to the benefits of of humane husbandry and proper land management, practices often at odds with ecological ideology.' *Country & Town House*

'Wondrous... Walling loves cows and I love his book. It intrigues and feeds just like a big, beautiful cow.' John Lewis-Stempel, *Countryfile*

Also by Philip Walling

Counting Sheep

TILL THE COWS COME HOME

The Story of Our Eternal Dependence

PHILIP WALLING

Atlantic Books
London

First published in Great Britain in 2018 by Atlantic Books,
an imprint of Atlantic Books Ltd.

This paperback edition first published in Great Britain in 2019
by Atlantic Books.

1 2 3 4 5 6 7 8 9

A CIP catalogue record for this book is available from
the British Library.

E-book ISBN: 978-1-78649-308-8
Paperback ISBN: 978-1-78649-307-1

The illustration credits on p.364 constitute an extension of
this copyright page.

Printed in Denmark by Nørhaven

Atlantic Books
An Imprint of Atlantic Books Ltd
Ormond House
26–27 Boswell Street
London
WC1N 3JZ

www.atlantic-books.co.uk

For Libby, who will understand better than most
what this book's about.

Contents

ACKNOWLEDGEMENTS

As with the writing of any book, various influences and events, over many decades, have contributed to its final form, and many of them I have either forgotten or have not been conscious of their effect. It is easier to remember the people who have directly helped me and I must record with thanks the great kindness shown by all those I visited and spoke to. If I have unintentionally forgotten anyone, I beg their forgiveness.

William and Richard Dart at Molland Botreaux and the Shinner family at Buckfastleigh in Devon; Jonathan Crump and Charles Martell in Gloucestershire; the Baynes family at Marleycote Walls and Hugh Richardson at Wheelbirks in Hexhamshire; Mark Gray at Broom House Farm in County Durham; and Don Antonio Miura at Lora del Río in Andalusia. I owe particular thanks to Simon Gray for his tremendous encouragement and friendship. I must also thank Charlie Bennett, in Northumberland, whose little herd of wild Dexters reminded me why my cousin advised me not to have anything to do with them.

Until I went to America and visited the natives in their natural habitat, I had no idea how generous and welcoming they are. It's a big country and it produces big-hearted people. I had never met any of them before I arrived in the US, but everyone treated me as if we had been friends all our lives. John and Rebecca Wampler spent two days showing me round the stockyards at Fort Worth, their ranch and their herd of Longhorn cattle. Rebecca was so generous with her time and enthusiasm, it is impossible to repay the debt; I can do no more than acknowledge how much I owe her. Russell Fairchild of the Longhorn Cattle Association, Gary and Kendra Rhodes in West, Texas, who made me feel so welcome, Kit Pharo and Tammy Fleischacker of the Pharo Cattle Company, with whom I spent a memorable summer day on the wide high plains of Colorado; the Yegerlehner family in Indiana, who impressed me with their quiet determination to farm against the American grain.

I regret having had no space to include an inspiring visit to Preston Correll in Kentucky, one of the most learned and intelligent men I have ever met; and a day spent with the wonderful Joel Salatin and his family and interns, in Virginia, whose often lonely evangelism for honest family farming has done much to show how the ravaged soils of North America can be healed.

My editor at Atlantic Books, James Nightingale, has been patience itself and I am grateful for his calm courtesy and wise advice. Jane Selley, my copy-editor, has done a sterling job, correcting many egregious mistakes; no doubt some will still remain, for which I take full responsibility.

Philip Walling
Scot's Gap
April 2018

Preface

W<small>E ARE THE</small> inheritors of a legacy of cattle breeding that stretches back into the ancient world. Without oxen to pull our ploughs and haul our carts, settled farming would have been impossible; without their manure, our soils would have been the poorer; without milk, our diet would have been deficient; and without their hides, we would have had no leather, or the myriad other things we take for granted. It is impossible to overstate the services rendered by the ox to the human race.

But humanity's dependence on cattle is not merely a thing of history. In many ways we are more reliant on them today than we ever were, although the form of our dependence has changed. As Western (particularly American) ways of eating have spread out into most corners of the globe, the demand for beef and milk has grown prodigiously. From modest beginnings in California in 1940, McDonald's now has 35,000 burger outlets worldwide. This American cultural and culinary influence has been achieved on the backs of the millions of cattle slaughtered annually to satisfy the apparently inexorable demand for minced beef in a bread bun – and, of course, the cow's milk that goes to make the processed cheese they put with it.

As this demand for what cattle produce increases everywhere, and eating steak is seen as a badge of affluence, so the numbers of cattle worldwide must only increase in step, from the current 1,000 million worldwide. India has 300 million, with 225 million in Brazil, 100 million in China, 90 million in the US, 90 million in the EU and even 18 million in Russia, not to mention the 10 million in the UK.

Most people in the urban West have little idea about cows. They cannot identify the breed, let alone whether it is a beef or dairy type. For example, they do not know that a cow has almost the same gestation period as a human female, or that some dairy cows can give 30 times their body weight in milk in a ten-month period. Or that a bullock of a specialized beef breed can grow to weigh a ton entirely from eating grass. This book is an attempt to give a flavour of what our cattle do for us. It does not purport to be more than impressionistic; more is omitted than included. It is certainly not an encyclopedia. There are plenty of those. It is simply an account of one man's recollection of his all-too-brief involvement with cattle.

I was 13 when my grandfather died and left me 50 acres of land, but I couldn't get my hands on it until I was 21. I wonder whether knowing that it was coming affected my attitude to schoolwork, and had something to do with my getting poor grades in my A levels. Who knows? Anyway, after messing about in a spoiled adolescent kind of way, and needing an occupation and an income, I bought an old Grey Fergie tractor and a tipping trailer for a few hundred pounds and started to do all kinds of jobs for anybody who would pay me. I accepted almost any work that was offered, but I found I had a particular talent for building dry-stone walls and got quite a bit of work. A few times I bit off more than I could chew and was nearly defeated by

2

a couple of bigger jobs, but I reluctantly came to accept my limitations.

One memorable job was building a wall around part of the graveyard of the Methodist chapel in the middle of Workington, just across the road from the bus station. This took me quite a long time. Every day I travelled the 12 miles there and 12 miles back on my old tractor (no cabs in those days), with the trailer loaded with sand and hundredweight bags of cement. I dug the foundations with a pick and shovel. In places my excavations were close to some gravestones, and one day I unearthed what I thought might be a human tibia – I remembered what they looked like from biology lessons. It was a bit creepy digging out half a decayed leg bone with a round joint on one end and a jagged break on the other. I wrapped it in a paper sack and stowed it under the tractor seat, intending to show it to the architect who was paying me to build the wall. Somewhere between Workington and home, however, it must have fallen off, because it was missing when I got back. I kept a careful lookout next morning on the road back to Workington, but I never saw it again. I've often wondered what happened to it.

During the summer, a neighbouring farmer asked me to help him for a few weeks. I had little idea about farming, but almost immediately I realized I had fallen in love with a world that had been closed to me even though I had been brought up alongside it. Here was adventure, pitting my physical strength against the land; working with my hands; the excitement of braving the weather; living with the changing seasons; driving tractors and machinery; working dogs, lambing ewes, calving cows and glorying in the earth's annual increase. I looked forward to every day and realized I had shed the melancholy that overwhelmed me as summer passed into the darkening days of autumn. I came to see why for pastoral people the new year is 1 November, when the rams go in with the ewes and the eternal annual cycle begins afresh. And autumn is sale time, when the year's increase turns into money and you feel secure to meet the winter with enough to see you through. It had more meaning than anything I'd been told at school.

Here was a world I could throw myself into; a world that ran along different lines from the one I'd been brought up in. I had stumbled upon a secret that had been there all along but that I'd never seen. Why were more people not desperate to get into this world? Why did people ever leave the land for those terraces of street houses I knew I could never have borne to live in?

There followed ten years of almost undimmed joy in my love affair with the land that passed almost in a flash. I can't remember where the time went. And with a lot of effort and a deal of good luck, I found myself owning my own farm. I kept beef and dairy cows, and had a milk round in the village; fattened lambs; kept hens and sold free-range eggs (until the trading standards people stopped me); reared and killed geese for the Christmas market; even kept a couple of pigs and made them into bacon and ham.

Then one day, when I turned 30, I gave it all up. I think it was the twice-daily grind of milking cows that got to me. Either that, or some gnawing inner voice prevailed and convinced me that I had missed out on an education and had to make up the deficiency while I was still young enough. I didn't know whether I would be clever enough to do a degree, but when I thought about the boys I had been at school with, most of whom had easily got into decent universities and were now doctors, lawyers or Indian chiefs, I couldn't believe I could be much less clever than they were.

One school friend told me he envied my success in having my own farm, whereas I saw it as coming to terms with academic failure. Another friend who had gone to Cambridge and become a solicitor couldn't understand why I wanted to give up farming. When I told him I rather fancied becoming a lawyer and farming on the side, he was incredulous: 'The law's just a job like any other; there's no magic to it!'

But what neither of them understood was that I felt I had missed out on an essential rite of passage that they had gone through and I hadn't. I needed that academic initiation to develop into adulthood, otherwise I would be stuck at an unformed stage for the rest of my life. I didn't want to give up the farm, however, and I had a vague idea that once I had been to university I would go back home and resume farming, maybe combined with practising law.

But once I got to university, I found myself on a path that led away from the land and from my Cumbrian roots to the bar and London and eventually to selling the farm. I suppose I wanted the money, but also I fondly thought I would always be able to buy it back, or another one like it, in the future. I had loved my time farming and had been good at it, but it was more than farming itself that tapped deep emotions; there were spiritual reasons that made it a vocation that matched my true nature. And in giving up farming and selling the farm, I came to feel that I had turned my back on something that really mattered to me. I had rebuffed that tutelary spirit the Romans called *genius* and the Greeks *daemon*; that personal guide that comes to everyone at birth and carries with it the fullness of our undeveloped powers. These it offers to us as we grow, and we can choose whether or not to accept and develop them in service to our *genius*, or turn our back on it.

Apuleius, the Roman author of *The Golden Ass*, wrote a treatise on the *genius/daemon*. He tells us that in Rome, on his birthday, it was traditional for a man to give something to his *genius*. In return, it would make him 'genial' – sexually potent, artistically creative and spiritually fertile. If a man cultivated through sacrifice and labour the gifts offered to him, his *genius* would eventually be liberated and, when he died, become a *lar*, a protective household god. But if a man ignored his *genius* because he felt no gratitude to it for its gifts, or he ignored its promptings, it would become a *larva* or a *lemur* when he died: a troublesome restless spirit left in bondage to prey on the living.

The farm was the most important thing in my life. My most prized possession. I can still close my eyes and visit every part of it. Doubtless the images are out of date, but I can see every hedge and wall, ditch and watering place, and every undulation in the land across every field on its 240 acres. I know where all the drains were, which were the walls with bad foundations that were liable to 'rush', the gates that swung and those that had to be lifted to open them. I had fenced the whole farm and laid most of the hedges while I was there, and rebuilt many of the dry-stone walls, so I knew the farm like a lover traces his love over every part of his beloved's body.

And then I threw it all away. Was it an act of great folly or a necessary sacrifice to my *genius/daemon* as I passed from one stage of life to another? The Celts understood sacrifice, both symbolic and actual. They practised it with determination, particularly in those liminal places at the threshold between this world and the next where the veil that divides the two is at its thinnest. Marshy places, neither land nor water, were places of transition where sacrificing their most precious possessions would have greatest effect. That is why Celtic artefacts, including magnificent swords, have been recovered from bogs and fens all across Europe.

In *Le Morte d'Arthur*, Excalibur, the fabulous sword given to King Arthur by the Lady of the Lake as a symbol of his status, had to be returned to the waters by Sir Bedivere at the direction of the dying king. Bedivere thought it 'sin and shame to cast away such a noble sword'. But his sovereign knew better. The sacrifice was necessary to mark his passage over the *limen*, the threshold, from life to death. And the lake represented a watery place of transition between this world and the next, where the sword could pass from the mortal knight to the immortal hand that rose from the waters to grasp it. One such beautifully forged sword, with jewelled and decorated pommel and now in the British Museum, was found at Embleton, in a boggy place, five miles from where I was born.

Since I parted with the farm and cast myself into exile from my roots in Cumberland, I have been trying to come to terms with my loss. Writing this book has been one way to attempt to get to the bottom of it all. Was it in defiance of my *genius*, or a necessary sacrifice as part of a rite of passage from one stage to the next? Why I couldn't have settled for a tattoo, as many people do nowadays, rather than giving away my most treasured possession, I will probably never understand. And if it was a sacrifice, has it done justice to my *genius*? Whatever it was, this book is written in the earnest hope that it will be acceptable as just gratitude for the gifts I received at birth and in proper homage to my tutelary spirit.

Introduction

Modern history has been much too sparing in its prose pictures of pastoral life. A great general or statesman has never lacked the love of a biographer; but the thoughts and labours of men who lived remote from cities, and silently built up an improved race of sheep or cattle, whose influence was to be felt in every market, have no adequate record.

H. H. Dixon (1869–1953)

'WHO KNOWS ANYTHING about wild flowers?' asked Mr Bacon at the start of my very first botany lesson at grammar school. We were outside the biology lab and he was holding up the flower on the end of a stem of silverweed.

I put up my hand enthusiastically.

'What is this called?'

'Silverweed, sir,' I replied.

'No. I want to know the name of this *part* of the flower.' And he pinched the pistil between his finger and thumb and held it in front of my face.

Over the previous five years, from the age of about seven, I had put together a considerable collection of pressed wild flowers, which I had pasted into three big scrapbooks. Wherever we travelled – and my parents were inveterate travellers all over Europe – I collected every wild flower I could

9

find, identified them and pressed them between sheets of blotting paper, which I kept underneath the mats in the car until we got home. The first thing I did when we stopped for a picnic was to scour the immediate area to see if there was a flower I hadn't got. I knew all their common names and could identify almost any wild plant on sight. My parents could do that for garden plants, but I had no interest in those.

It wasn't a scientific interest. I was drawn to their beauty – and their names. Not their Latin names, but their traditional English ones. I didn't care what the parts of the plant were called, or what they did. And I didn't know, when Mr Bacon asked me to identify the constituent parts of a flower, that it was important.

'What's this part called?' he asked, this time lifting a sepal with his forefinger. 'Silverweed, sir!' shouted out a sharp little boy whose name I didn't yet know, but who would later become a comrade in rebellion. The whole class laughed and I was mortified.

I think that was the day I lost my passion for pressing flowers. I didn't then know that it would be transmuted into something else – delight for the land.

It's been a long time since ordinary people in industrial countries have had much idea where their food comes from. They've become increasingly isolated from its production and increasingly disdainful of the people who produce it. But it wasn't always like this. At the beginning of the Industrial Revolution in Britain, farmers were valued as practical, intelligent and enterprising men who rose to the challenge of feeding an ever-expanding nation. They transformed England's acres into some of the most productive land in Europe, despite the often unfavourable climate and poverty of the soil.

Farming under George III – 'Farmer George' – became the pursuit of kings. Landowners and gentry threw themselves enthusiastically into breeding livestock and farming, and granted secure tenancies on their estates to able and improving tenants on long leases protected under the common law, good against the whole world, including the landlord.

This, and more, meant that as the Industrial Revolution gathered pace and the population burgeoned, British farming took on the task of feeding the nation. As the British climate and terrain is more suited to growing grass than almost any other plant, it was livestock farming, particularly with cattle, that turned that grass into the energy and productivity needed.

It was not only food that the nation required in ever-increasing quantities. It also needed leather, for a vast range of essential uses from horse harness to pulley belts; tallow (candles were the main source of light); and oxen and later horses for nearly all the motive power. Without the concurrent revolution in farming, Britain would never have been able to turn itself into the first global powerhouse, whose ideas and inventions spread across the world. And at the heart of this were cattle. Without the fertility created by cow muck (farmyard manure, as it's known euphemistically; FYM for short), none of it would have been possible.

Cattle are noble animals and their keeping is a noble endeavour. But in tune with the strange topsy-turvy world we currently inhabit, we have come to disdain the domestic animals upon which we depend and instead to revere the 'natural' world. We mostly treat our domestic animals with indifference, denying the extent of our dependence and consigning them to short, utilitarian, industrial lives, scientifically fed and efficiently slaughtered out of sight. We cannot reconcile their beauty with their necessary killing, so we no longer mark the death of an animal with ritual or thanksgiving, but keep it hidden, as if we are ashamed. We seem to find it hard to be capable of respect and gratitude for the lives of animals that have been given to us as a gift. Ironically, one of the few uses to which we put cattle where we do treat them with the respect due is the bullfight. Every bull is given a name, his life and death in the arena are dedicated to someone important to the matador, and sometimes his head is preserved and displayed.

Until recently, most domestic cattle had a triple purpose: traction, milk and meat (at the end of their lives), as well as all the other uses to which their carcases were put when they were dead. But as cattle have been forced to be more productive, the breeds have become more specialized and are now divided into either beef or dairy, depending on the animal's dominant purpose, although there is some crossover because the males bred from dairy cows do supply a large proportion of our beef. There is a third category that are neither dairy nor beef, but are kept for some other purpose, even though they may well be eaten at the end; these include parkland cattle, such as the Chillingham; fighting bulls; Texas Longhorns, now kept for the length of their horn; and Heck cattle, which are bred to try to recreate the aurochs, in order to prove that domestic cattle came from a primitive ancestor. There are also those millions of cattle in India that will never be eaten because they are seen as avatars of a deity and protected by law from harm.

Wherever humans have migrated they have taken their cattle with them. They are our longest-serving domestic animals, which since the beginning have tilled our soil, borne our burdens, fed us, clothed us and been our uncomplaining servants in the work of taming the wilderness and wresting a living from it. There has never been a time when we have not depended on cattle. And even though many people have migrated from the land to the city and the things we need from cattle have changed, that fundamental dependence remains. Yet as people both in the West and across the developing world retreat further into a virtual world, isolated from the real one, where fantasy becomes reality generated by computers and electronic games, they come into contact less and less with the life of the land that underpins everything they depend on. In the space of three or four generations, we have lost touch with the source of our food.

Cattle are our oldest form of wealth. Amongst many central African tribes, they are still currency. In ancient Greece, even after metal coinage superseded cattle as a means of exchange, the image of an ox was stamped on the new coins to give them authenticity. The Latin for cattle is *pecus* (Proto-Indo-European *peku* and Sanskrit *pasu*), which gives us 'pecuniary'. In Old Saxon, cattle was *fehu* (with the same root), from which we get 'fee'. The English word 'cattle' was borrowed from the Norman French *catel*, which had come via medieval Latin *capitale*, 'principal sum of money or capital', from the Latin *caput* meaning 'head'. Moveable assets are still described in English law as chattels. The word 'cow' came via the Anglo-Saxon *cū*, from an Indo-European word *gōus* meaning a bovine animal. The plural, *cȳ*, became *ki* or *kie* in Middle English, and then acquired an added plural ending, giving the Old English *kine*.

So where did they come from, these creatures upon which we depend so much? After Darwin, it became accepted wisdom that domestic European cattle – *Bos taurus* – are descended from the aurochs, the wild ox, *Bos primigenius primigenius*. Even its Latin taxonomic name, *primigenius*, the firstborn, supposes it to have been the original, the precursor, to the later *Bos taurus*. The words aurochs, urus and wisent have, in the past, been used interchangeably in English to describe this wild ox. The Romans borrowed the Germanic word *ūr*, compounded with *ohso* ('ox'), giving *ūrohso*, to make *urus* (plural *uri*) in Latin to describe the wild beasts they found in central Europe. This then was borrowed back to become *Aurochs* in early modern German (singular and plural, meaning 'primeval ox' or 'proto-ox'). It is directly parallel to the German plural *Ochsen* (singular *Ochse*) and echoes the English words 'ox' and 'oxen'.

In his *Gallic Wars*, Julius Caesar describes the wild cattle he encountered in central Europe (with perhaps a touch of literary licence) as being 'a little below the elephant in size, and of the appearance, colour, and shape of a bull. Their strength and speed are extraordinary; they spare neither man nor wild beast which they have espied.' When Caesar saw these formidable beasts, they had existed for millennia; their remains (fragments of bone and horn) have been found from the Pliocene (5.3 million to 2.5 million years ago) into the Pleistocene period (which ended about 11,500 years ago), tending to show that their range expanded and contracted as the climate grew warmer and colder with various climatic cycles. As the world entered its current warm period and the last ice retreated, these ferocious ruminants proliferated across almost the whole European landmass into Asia and

Siberia, except northern Scandinavia and northern parts of Russia and Ireland.

The last aurochs cow is said to have died in a wood in Poland in 1627. It was a separate species from the extant wisent, the European bison. The two have often been confused in the past, and some sixteenth-century illustrations show aurochs and wisents with similar features. People were less concerned with verisimilitude than the fact that these were large, dangerous and terrifying beasts of the European wildwood.

It was because they appear to have been so ubiquitous in the wild state that it has been assumed that the aurochs must have been the ancestors of at least some of our domesticated European cattle. But recent research disavows widespread European domestication and says it started with a few wild cattle in the Near East tamed by some of the first settled farmers, who took their cattle with them as they spread outwards into Europe and Asia.

Tracing their Levantine origin takes us to the Fertile Crescent, the 'cradle of civilization' that lies in a great arc from the lower valley of the Nile, up the eastern seaboard of the Mediterranean and sweeping round to encompass the Tigris and Euphrates rivers down to their delta in the Persian Gulf. In the late 1940s and early 1950s, on the eastern edge of this region, at Jarmo, in the Zagros Mountains (now in Iraqi Kurdistan), Robert Braidwood (1907–2003) found evidence of farming with domestic livestock dating from about 7000 BC. Digging unearthed what were believed to have been cattle enclosures, complete with bones. Despite the best modern analysis, it is not clear whether these bones are of domestic cattle or what might have been their wild ancestors. Researchers have made assumptions from the size and age of the sample – the bigger and older the bone, the wilder the animal is assumed to have been – but they have been unable to determine what exactly was grazing the land all those years ago.

However, in excavated kitchen debris Braidwood found that 95 per cent of the bones were of domestic animals – sheep, goats, cattle, pigs and dogs – while only 5 per cent were of wild animals. More recent digging at Chogha Golan, in the same Zagros Mountains, has revealed evidence of agriculture going back even further, to about 12,000 BC, showing that people had domestic cattle at least by this date – only a thousand years or so after the earth began to warm after the last ice retreated. It supports scholarly agreement that agriculture originated here and spread out across Europe and Asia over the ensuing millennia.

The two sites where the earliest remains of what are believed to be domesticated cattle have been found, Dja'de and Çayönü, are less than 150 miles apart, which suggests a limited area of domestication in the Levant. Recent mitochondrial DNA analysis (descent from mother to child of both sexes) supports this theory that there was little or no wider domestication of western European aurochs. Yet more recent research* tends to the view that there were very few cattle originally involved – at most 80 and probably fewer – and that these are the ancestors of all European cattle. If that is right, then every living cow will be descended from those original cows and no others.

* R. Bollongino et al., 'Modern Taurine Cattle Descended from Small Number of Near-Eastern Founders', *Oxford Journal of Molecular Biology and Evolution* (September 2012).

Other researchers will not concede this and assume there must have been other attempts at domestication across a wider area: early farmers must have tried to domesticate aurochs because they are believed to have been so ubiquitous. When it is pointed out that there is little or no evidence of *any* domestication of wild oxen in other places, they reply that early farmers could have tried and failed because the aurochs would have been too difficult to handle. This is a fine example of researchers looking for evidence to support a theory and not the other way round.

Recent work (2013) by Arne Ludwig at the Leibniz Institute in Berlin and Lawrence Alderson, a founder member of the Rare Breeds Survival Trust (and others), has traced the matrilineal descent of a White Park cow in Alderson's Dynevor herd to a cow that lived in Britain 10,000 years ago. This tends to tell us that we should take a long view, not only of the length of time we have been keeping cattle in Britain, but also of the period over which they have been domesticated. In a word, nobody really knows. Results depend on whether researchers follow a haplogroup (a group of genes inherited from a single parent) through Y chromosomes (father to son) or through mitochondrial DNA (from mother to child of both sexes).

If we assume there was no domestication of aurochs in Europe, then the bulls painted 17,500 years ago on the walls of the caves at Lascaux must have been wild beasts that our ancestors hunted, and not the forebears of domestic cattle. And the paintings were probably done for the same reason as hunters today have themselves photographed with the heads of the game they have shot and mounted on their walls. Perhaps if our Neolithic ancestors had known the art of taxidermy, and their specimens had survived, we would have had a better record of what they hunted. Instead we must be content with those cave paintings that have only survived because of a happy accident of topography and climate.

There is another, older explanation for the origin of our cattle, which is rather outmoded now but which, by what might be more than a coincidence, leads us just up the road from the Zagros Mountains, to Mount Ararat, where Noah's Ark came to rest after the Flood. Only a few years before Darwin published *The Origin of Species*, William Youatt, in the introduction to his great work *Cattle* (1834), traced the 'native country of the ox, reckoning from the time of the flood … to the plains of Ararat, and he was a domesticated animal when he issued from the ark. He was found wherever the sons of Noah migrated, for he was necessary for the existence of man; and even to the present day, wherever man has trodden, he is found in a domesticated or wild state.'

Genesis tells us that our domestic cattle had been our necessary and constant companions since the creation of the world. Even before the Flood, Jabal, the son of Lamech (probably born during the lifetime of Adam), was 'the father of such as dwell in tents, and of such as have cattle'. After the Flood, there are numerous biblical references to cattle forming part of the wealth of individuals. When Abraham went into Egypt to escape famine, Pharaoh gave him sheep and oxen and he became 'very rich in cattle'. Both he and Lot, his nephew, owned so many cattle that they had 'herdmen' to look after them. These were almost certainly domestic cattle, because they are being looked after in herds.

The Bible is not the only source for these kinds of references. In Hindu scripture, the domestic cow was the first living thing created by the supreme deity after mankind, and as a result people are forbidden to harm it or shed its blood. Cows were not domesticated from wild species, but given by God as our servants, a separate order of creation from wild creatures. Out of gratitude for God's gift they were to be treated with reverence; for example, neither Hindus nor Jews were allowed to muzzle an ox when it was threshing corn. Even the bloodthirsty Romans punished by exile anyone wantonly killing an ox, although that was more to do with its usefulness than out of any humanitarian concern for its welfare.

It is suggested that we should not be too quick to reject the biblical narrative, because when its modern rival is examined critically, it makes hardly any more sense and begs as many questions as it answers. Domestication is a hard thing to achieve. It is not the same as 'taming'. Any wild animal can be taken when young enough and brought up away from its parents until it reaches adulthood. As a juvenile it will respond to its keepers in a broadly similar way to how it would respond to its natural parents. But once it is mature, its true nature will emerge and it will tend to behave according to its genetic inheritance, unless in some way its nature has been modified so that it is capable of being enfolded into human society, and that modification has become fixed and heritable.

Any scientific hypothesis that depends on our ancestors domesticating a few wild cows is also hard to accept on a practical, common-sense level. Have any of these scientists ever handled a difficult bull, like a wild Dexter, or encountered a Spanish fighting bull at close quarters? Yet these are both domesticated. At some point one wild beast of the kind that Caesar describes, violently hostile to mankind, must actually have been used to start the process. It can't have been done over a short period of time, like training a dog, because it involves transforming its nature, or that of its offspring, from ferocious to docile. You can hand-rear an alligator from an egg and let it swim in your bath, but you will never turn it into a trustworthy domestic pet. And if that's too extreme an example, consider the dingo. In certain states in Australia it is still illegal to keep one or to cross it with a domestic dog, because there is something in its nature that no amount of human handling can reliably domesticate.

Also it is by no means clear which came first. Did the domesticated ox give the first farmers the wherewithal to settle to farming; or, having determined to give up hunter-gathering, did they then domesticate the ox for the multiple purposes for which they needed it? And if they domesticated it first, why did they do it? If they had never seen such an animal before, how can they have known what they were looking for, or that it might have been possible to create an animal (from the ferocious type Caesar described) that would be docile enough to train to the cart and plough? And if they settled down to farming first, how did they till the soil and cart its produce before they had domesticated an ox? It is easy to imagine that at some point one or more aurochs bulls might have bred with already domesticated cows – it can't have been the other way round, because the calves would have been reared with the wild herd and lost to domestica-tion. But that does not answer the question of where they found the domestic cows in the first place.

I had enough trouble getting some newly calved Friesian heifers to stand still long enough not to kick off the teat cups when I first tried to milk them. With one in particular,

I had to loop a piece of rope around her belly in a slip knot just in front of the udder and over her back, and pull it as tight as I could in front of the hip bone, so that she was nipped up in two segments, like a caterpillar. This stopped her from kicking out sideways. I also used an aluminium device with a crook at each end. One end went into the groin and the other pressed on the nerve near the hip bone. This was not as effective as the rope, but it was quicker to put on.

And this was an animal with 12,000 years of domestication behind it. I doubt our Neolithic forebears could have got within half a mile of an aurochs, let alone restrained one with a rope. Did they even have ropes in Neolithic times?

Derek Gow has experience of trying to work with the modern version of the aurochs. In 2009, he brought 13 Heck cattle from Belgium to his farm in Devon. These were descended from cattle bred in Germany at the beginning of the Weimar Republic in the 1920s and 30s by the brothers Heinz and Lutz Heck. Hermann Goering and the Nazis were ardent environmentalists, and supported the Heck brothers' efforts to recreate the long-extinct aurochs by back-crossing breeds they thought might be their nearest living descendants. Lutz Heck was director of the Berlin zoological gardens and Heinz of the Munich equivalent. It took them 11 years to achieve what they believed to be success, each of them using different breeds. Lutz included Spanish fighting bulls in his melange, while Heinz preferred other varieties, one of which was wild Corsican cattle. None of Lutz's cattle in the Berlin zoological gardens survived the Soviet occupation after 1945, but some of Heinz's fierce throwbacks lived on as the ancestors of the modern Heck cattle.

Their breeding has been criticized as no more than fancy because they are just domestic cattle with a bit of wildness bred into them. They have also been denigrated as 'Nazi cattle', and it has been said that they do not resemble the aurochs – although their detractors can have little idea of what the true aurochs looked like. It is argued that the ancient aurochs would have been much bigger, fiercer, more muscular and with longer legs, and the modern version has no more of these characteristics than do, for example, the Spanish fighting cattle.

Gow's 13 Heck cattle grew into a herd of over 20, which, like that other product of scientific experimentation, Frankenstein's monster, ran out of control. By January 2015, he had to shoot all but six of his herd because they were simply too dangerous to handle. It is worth remembering that all the ancestors of these cattle had been through a process of domestication. None of them was a wild beast when the Heck brothers used them in their experiments to try to reverse domestication. Although they made them wild enough to be dangerous, it proved impossible to take domestic animals and breed the wild back into them. How much harder would it have been, without all the modern facilities, to take even one wild beast and domesticate it.

Until genetic evidence can be found to show that our domestic cattle have *some* aurochs DNA, it seems the best that can be said is that we have had domestic cattle for at least 10,000 years, not descended, but as a separate species from the wild variety. And that takes us back to the beginning of the Neolithic period, when we are told that people made the transition from hunter-gathering to settled farming. But as further evidence comes to light, and we find we are having to extend back into 'pre-history', the beginning of human agricultural settlement, it must follow that our domestic cattle, being at least as old as farming, have been with us for a very long time indeed. Where they came from I do not know, but as things stand, neither does anybody else. It pleases me to believe that we have had them as long as we have been human, as our constant companions and partners in the great endeavour of taming the wilderness.

For sheer dogged power, the ox could not be beaten. Pound for pound it was stronger, had more stamina, was cheaper to keep because it was a better converter than the horse of poor-quality roughage into energy, and was less demanding to look after. And it could be eaten when its time was up, without any of the revulsion and guilt that our society associates with consuming horseflesh. As oxen gradually gave way around 200 years ago to horses, partly for reasons of speed, cattle lost their ancient triple function and their breeding resolved itself into two distinct purposes. Just as sheep breeders had to decide between wool and meat, because it is impossible to improve the carcase of a sheep without the wool deteriorating, so the impossibility of improving the milk yield of a cow without its meat suffering caused cattle to become either dairy or beef producers. During this transition into specialization, many breeds retained their function of dual-purpose producers of milk and meat, but as the nineteenth century progressed, different breeds took separate paths. And in the last half-century, these differences have become ever more pronounced.

This book is an attempt to give a flavour of the very long road that our faithful cattle have trodden with us. For throughout history, whenever and wherever in the world we have needed their strength, their manure, their milk, their meat, and all that we get from their carcases when they are dead, our ever-dependable, uncomplaining oxen have been at our side, enriching our lives.

CHAPTER I

Dairying

Believing agriculture to be well calculated to improve the virtue, and call forth the talents of men, I have taken every opportunity of showing its superiority to all other pursuits.

William Aiton (1731–93)

MILKING COWS IS a special form of slavery. The responsibility is relentless. Dairy cows, like all animals, are creatures of habit and they give of their best if they have a routine. They have to be milked at the same time, twice (sometimes three times) a day. Every morning the first thing you do is get up and milk the cows. It doesn't matter what else is happening that morning; the cows have to be milked before breakfast. And whatever you are doing, wherever you are, you have to be back home by five o'clock for the milking, and unless you have help, you can't take a single day off.

I had 60 milking cows, which took an hour to milk if I got a move on. I milked them in a byre converted into a milking parlour. Even back in the early 1980s, when milk fetched quite a good price, there wasn't enough profit in it to employ anybody to help me. Every morning of the year,

Christmas Day included, at half past eight, give or take five minutes, the milk tanker came to pick up the milk from that morning's and the previous evening's milking. It had to be cooled, otherwise the driver could refuse to take it. To get it down to the right temperature took about three quarters of an hour, so the milking had to be done by around 7.45, which meant that I had to start about 6.30.

On a summer morning, with the early sun creeping down the fellside and warming the still air, it was pure joy to plunge outside into the new day. But pulling on my overalls, stiff with cow muck, and dragging myself out of the house at six o'clock on a pitch-dark January morning, with freezing rain lashing the bedroom window, was less of a pleasure. Rolling over was out of the question; I simply could not fail to milk those cows. Illness had to be ignored because there was nobody else who could do the milking. There was no point in even allowing myself to admit to having flu, or a thumping hangover, or, on one occasion, food poisoning, because it just made the task even harder. Only once, when I was too ill to do it after I had suffered a welding flash, did my wife ask my neighbour from down the valley to milk the cows.

Once I got started, the milking had a way of creating its own momentum and I would lose myself in the mechanical repetition. I became focused on getting through it as fast as I could and tended to ignore everything but the diurnal work of keeping a dairy herd, with record-keeping and planning ahead falling by the wayside. I was particularly bad at recording when each cow had calved and when she had been served by the bull or artificial insemination (AI). And if I didn't know how long a cow had been milking, I didn't know when she ought to be dried off to give her a rest to prepare for her next calving and lactation.

Some farmers are suited to the routine and certainty of milking cows. They accept it as something that has to be done before the work of the day begins. And at one time it paid the bills, put a bottom in a farming business and was the financial salvation of many small farmers. But paradoxically, for many others, especially if they were one-man bands, it proved to be their nemesis. Milking became the focus of their day's activity; once it had been done, they felt they could take it easy until the afternoon because they had made their money for the day. From being only one part of a well-run farm, the dairy herd began to consume most of their effort and attention.

For about ten years, maybe longer, after I gave up milking, I suffered a series of nightmares about the whole process. In fact I still have them now and again, but not as intensely as I once did. In one, I have forgotten to milk the cows for a long time – many days, maybe a week or more – and they are locked in their shed, where they haven't been fed. Some have split and burst udders; others are lying in a khaki mixture of milk and their own liquid excrement, bloated and unable to get to their feet. Some are standing in agony with grotesquely swollen udders, milk streaming from their teats. Some are dead; one looks as if she died trying to give birth, a pair of hooves and a hideous head with grotesquely swollen and blackened tongue flopping from her vagina; pink froth like ectoplasm is congealed around its nostrils and cold dead muzzle. I am revolted by my criminal negligence and I do not know how to put it right.

In another dream I am milking eight cows side by side in the byre. I have attached the pulsing rubber cups, one to each of their four teats, and I can see the milk through the little glass inserts, coursing down the tubes up into the receiving jars. The dogs start barking and I go outside into the yard to investigate. A delivery van has arrived with a parcel to sign for. I take the parcel, sign the sheet and have a chat with the driver. When I get back to the cows, I find that time has played a trick on me and I have been away for more than an hour. During that time, the milking machines have milked the cows dry. Three of them are being sucked down into the teat cups; half of one cow has disappeared and the udders and bellies of the other two have gone. I rush to pull off the cups, but they have such a strong hold that I cannot remove them. As I am struggling with the first three, another cow is being sucked through the teat cups and down the milk line. The suction is too strong and the other cows are being hoovered up bit by bit to the rhythmic ker-plop ... ker-plop ... ker-plop of the milking machine, and there is nothing I can do to stop it.

I wake up in a desperate state. How could I have allowed this to happen? I am weighed down by terrible guilt that my neglect has caused such suffering to animals in my care, and I only gradually realize that it was a dream. Just as hearing the voice of his master after a slave has gained his freedom penetrates his soul and takes him back into a horror he can never leave behind, for many years hearing the sound of a milking machine took me straight back to the misery of that milking parlour all those years ago.

I'm not alone. Milking cows has driven many men mad.

One less than energetic local farmer, who didn't enjoy early mornings or repairing his fences, or the work of mucking out his little herd of cattle, let them wander at will all winter over the whole farm. On dark mornings, when they were hungry, the herd would make their way into the farmyard and mill about, mooing, waiting to be fed. He solved this hindrance to his slumbers by carrying bales of hay upstairs and stacking them in his bedroom beside the window. When the cattle turned up for breakfast, he would throw up the sash window, cut the string on the bales and toss out the sections of the hay bales, frisbee-like, to the hungry herd before going back to bed.

After his wife died, standards slipped even further. He lived in his wellingtons and bib-and-brace overalls, indoors and out. Decades of *Farmers Weekly*, *Farmers Guardian*, *British Farmer and Stockbreeder* and the local paper were stacked up in the hallway. A full-grown sheep that had once been a pet lamb roamed at large through the house. One afternoon I was invited in for tea. The kitchen was carpeted with flattened cardboard boxes. When they got too dirty, the farmer simply gathered them up, burned them and replaced them with fresh ones from the supermarket where he bought his groceries. I sat at the kitchen table while he produced two tea-stained chipped mugs and filled them from a teapot he kept warm on the back of the Aga. He dropped a half-empty bag of sugar on the table.

'Sugar?'

Before I could answer, the sheep had got its muzzle into the bag and was gobbling at the contents. The farmer recognized the rustling and without turning round said:

'Just knock that yow's head out of the bag if you want any.'

What saved many small family farmers from penury was the Milk Marketing Board. Formed in 1933 to protect them against market instability and the dominance of wholesale buyers, it put a bottom in the market after farm commodity prices collapsed following the First World War, leaving dairy farmers in dire straits. The MMB was legally obliged to buy every gallon of milk produced by English farmers. The other countries in the UK had corresponding boards. It rapidly became the largest milk marketing and processing organization in the world, which at its peak bought and marketed 13,000 million litres of liquid milk a year.

During the Second World War, the Ministry of Food tightened the state's control and made the MMB the direct buyer and seller of the nation's milk. It also ran the National Milk Records service, under which it provided milk record-ers to visit farms weekly to record the yield of each cow. This gave farmers the impetus to keep accurate records, which greatly increased production through higher-yielding cows.

At its height, the MMB employed 7,000 people in four separate businesses: the Milk Marketing Scheme, which promoted the sale of milk; Dairy Crest, through which it collected, processed and delivered milk; its Genus AI service; and National Milk Records.

After the war, the MMB continued to guarantee the price for all liquid milk, and fixed the retail and wholesale prices in consultation with buyers. By the 1950s, production had increased considerably, as many small farmers took advantage of the guaranteed income. When refrigeration and bulk tank collection reduced the labour of collecting the milk in churns from the roadside, production was boosted still further.

By 1961, a quota system was proposed to regulate the ever-increasing supply, but the MMB rejected it. As Britain negotiated to join the Common Market, it became clear to a few people that the MMB could not co-exist with European law and the European system of dairy cooperatives and import controls. Sooner or later the British model of price guarantees without import restrictions would have to go.

Rather than frighten farmers, however, and turn them against joining the Common Market, the European Commission allowed the MMB to keep its statutory monopoly until 1978 by exceptionally allowing the Board to negotiate amendments to the Common Market regulations. But the wolves were circling, and inevitably, in 1982, the MMB lost a challenge to its pricing structure by the European Commission in the European Court, and an Irish dairy company obtained damages from it for unfair competition under EEC rules.

Further EEC regulations attacked the MMB's control of liquid milk prices and purchasing from one flank, while the supermarkets attacked from the other. The wounded MMB limped on, until in 1987 the European Court decided that the profitable Dairy Crest had to be separated from the MMB because it was contrary to European competition law. It became clear that the Board was doomed. The Conservative government in the late 1980s could do nothing to prevent its abolition, and in 1994, the minister of agriculture, John Selwyn Gummer (as he then was) announced the *coup de grâce*, disingenuously pretending that it was a British decision to privatize the milk market and 'free it' from the MMB's monopoly.

The original European system of price support for farmers was based on 'intervention buying', by which the Commission bought produce off the market if the price fell below that which had been set for it. It was then to be stored until the price rose again, when it could be fed back into the market, causing the price to fall back to the level set. It was a typically utopian and dirigiste scheme that simply did not work in practice. It caused huge accumulations of dairy produce in 'milk lakes' and 'butter mountains', which the Commission could never feed back into the market because production kept rising. They ended up virtually giving it away, most famously to Russia and 'developing' countries. In turn, this dumping of cheap produce distorted local markets and damaged small farmers trying to make a living in these countries.

In 1984, the Commission came up with a bureaucratic and byzantine milk quota system designed to regulate the over-supply. Member states were set a national limit on dairy production; they in turn set a quota for every farmer who had been producing milk in 1981 based on his share of the national quota plus 5 per cent. Member states were fined if they exceeded their national quota, and each state imposed levies on its individual farmers if they exceeded their personal quota. But an individual farmer would avoid a penalty if the overall national production was less than a member state's allocation, because the EU would not impose a national penalty. The catch was that this was based on a previous year's national production, so farmers could never know at the time they sold their milk whether or not it would attract a levy. Fortune tended to favour the bold, and often, if a farmer took the risk that national production would not exceed the EU limit, he got away with going beyond his individual quota. Those who stuck honestly to their allotment lost out.

This was an unsatisfactory and risky way to run a business. Prudent farmers who knew they were going to exceed their quota found it safer to buy or lease spare quota in the market that had sprung up. The EU had made quota transferable only with the land to which it was attached, but in practice a short grazing licence was enough to legitimize the transaction. The buyer would rent some land with quota attached for a season and use the quota to sell that amount of milk to the wholesale dairy. Then at the end of the licence period he would simply vacate the land and neglect to transfer the quota back to the registered holder. The 'rent' was set by the value of the quota in the quota market. By this method, nicknamed 'quota massaging', large quantities of quota changed hands and became detached from the land to which it had originally been granted. This is an example of the policies of the European Commission leading to results it was unable to foresee.

Quotas had to go, and in March 2015 the EU abruptly abolished them. The Commission justified it by claiming that demand for dairy produce, particularly cheese, had increased in Europe and across Asia and quotas were no longer needed, even though European farmers still gener-ated more dairy produce than was consumed in Europe. The truth was that there was less incentive for farmers to increase production because they were no longer guaranteed the kind of prices that caused overproduction. The EU had turned to payments to support farmers 'decoupled' from production: in other words paying them for doing things on their land other than producing food, such as environmental schemes and social payments to sustain rural communities.

All this has seen a drastic reduction in the number of farmers milking cows in the UK – and to a lesser extent across Europe. In 1950, there were 200,000 individ-ual farmers delivering milk to the MMB; by November

2016, this number had fallen to 9,500. The national herd decreased from 2.5 million cows in 1990 to 1.8 million in 2015, while the milk produced increased by a third. With a few peaks and troughs, the wholesale price of liquid milk has hardly altered in a decade and is now little more than it was six years ago – about 31p a litre, a few pence a litre less than the average cost of production. The price in mainland Europe is generally higher, particularly in Holland at 37p and Germany at 36p a litre. This goes some way to explain why British farmers find it hard to compete.

The loss of price support turned out to be a catastrophe that nearly overwhelmed farmers and brought the dairy industry to the verge of collapse. The market returned to its pre-MMB volatility. From 2000, the milk price fell 40 per cent in 18 months, and with it the value of cattle. Supermarkets quickly seized the power to squeeze the wholesale price below farmers' cost of production and carried on charging their customers more than twice what they paid for it; in most cases they were selling milk more cheaply than bottled water.

Slowly it has begun to dawn on farmers (the more far-sighted realized it a long time ago) that they are on their own and cannot look to their government to support them. If they want to continue milking cows, there are only two ways of doing it: escape the roller coaster of world markets and the stranglehold of dairy processors by taking control and selling direct to the customer; or borrow money, enlarge the herd and turn the farm into an industrial unit, spreading the cost over hundreds of ultra-high-yielding black-and-white Holsteins, the industrial cow of choice across the world for farmers chasing yield. But first, an ancient English dairy breed with a modern purpose.

CHAPTER 2

The Gloucester Cow

O N 10 OCTOBER 732, somewhere between Tours and Poitiers, a Frankish army led by Charles Martel defeated a Moorish horde under Abdul Rahman Al Ghafiqi, the Muslim governor of Spain. Intent on conquering Gaul for Islam, Rahman had crossed the Pyrenees, defeated Odo, Duke of Aquitaine, and advanced as far as the Loire, pillaging and burning as he went. Exactly a hundred years after the death of Muhammad, this was the culmination of two decades of seemingly unstoppable Muslim conquest of huge swathes of territory across North Africa, the Middle East and into southern Europe.

The Battle of Poitiers is arguably the most significant event in the history of the world, certainly of Europe, because the survival of Christendom hung on its outcome. Had Charles Martel not triumphed over the Moors, Islam would very likely have prevailed throughout Europe. There would have been no Charlemagne, no Holy Roman Empire; 1,500 years of Christian Europe would never have been and the history of the world would have been utterly different.

The modern Charles Martell, from Dymock in Gloucestershire, is reticent about admitting that he might be of the same lineage, but playing a crucial role in saving the Gloucester cow from extinction, as well as preserving hundreds of old varieties of apples and pears and inventing Stinking Bishop cheese, might show that some of the blood of the saviour of Christendom has trickled down the generations.

The present Charles Martell settled in Gloucestershire, near the border with Herefordshire, as a young man with little money, his heart set on farming a piece of his own land. His grandfather had left him a cottage, which he sold for £15,000, and in October 1972, with the money in his pocket, he came to Laurel Farm, Brooms Green, where he stood in the farmyard and made the winning bid at the auction. You could do that in those days. There were hundreds of small farms sold by auction up and down the land for a price commensurate with the profit you could make from them, which allowed young men with enterprise to get into farming. People weren't buying farms then to split them up and sell off the buildings, or for the subsidies that came with the land for doing nothing.

After Martell had acquired his few acres of fields and orchards and the run-down house and buildings, he had no money left. With a wife and two small children to support, and no income, he found work driving a lorry, sometimes walking the five miles to the depot to start his day and back home again in the evening. Losing the job after a year turned out to be a godsend, because it propelled him into selling cheese at markets around Somerset, Gloucestershire and beyond. He soon realized he could make more in a day from his cheese stall than he'd earned in a week driving lorries. And that gave him the start he needed.

He might have understood the principle of livestock-keeping, expressed in William Youatt's great aperçu that 'the grand secret is to match the breed to the soil and the climate'. Or he might just have been an idealistic romantic. Many young people were back then; it came with hippies and flower power. I was one myself for a couple of years until I woke up to reality. Whichever it was, he thought native Gloucester cows would be the right breed for his little farm: he would be saving an almost extinct breed, and it would give his farming some purpose beyond the merely commercial and utilitarian.

But there was a more hard-headed reason too, otherwise he would not have survived. Gloucester milk makes good cheese, better than milk from most other breeds because of the fine balance and quality of the fat and protein. The fat globules are small, so the curds separate more easily from the whey, and the greater milk solids means that more cheese can be made from each litre than from ordinary milk. That is why milk yield is recorded by weight rather than volume. A gallon of cow's milk weighs between 8¼ and 8¾ lb, depending on the breed and the way the cow is fed. Gloucester milk contains between 5 and 6 per cent butterfat, and solids-not-fat (protein and milk sugars) of 8–9 per cent. By contrast, the milk from high-yielding Holstein cows has about 3.5 per cent fat and about 8 per cent solids-not-fat. The rest of the milk is water. And once the cream is skimmed off, it's over 90 per cent water.

Broadly speaking, the dairying counties of England overlie a band of Jurassic Lower Lias limestone that stretches from the North York Moors through Lincolnshire, Leicestershire, Somerset and Dorset, with outliers in Cheshire and

Glamorganshire. Stilton from Leicestershire and Cheddar from Somerset are probably the best-known English cheeses, although Cheshire, Caerphilly and Gloucester are not far behind. All these counties are on limestone and have been renowned for their cheese for centuries.

In each of these places distinctive types of cattle evolved that could make best use of the terroir and produce milk that was suitable for cheese-making. They were not breeds as we know them today; rather they were types that had acquired a degree of genetic uniformity influenced by topography and climate, much more the products of the soil they lived on than modern breeds of cattle. Most of them would be the offspring of a bull that had only walked from the neighbouring parish rather than been flown from the other side of the world as a straw of semen. Modern cattle are generally mated according to measured performance, but these types were created using the observations of breeders who had little to go on other than evident secondary characteristics such as the thickness of a cow's 'milk vein', or the width of her muzzle, or the colour of her hooves or coat. Not that these were necessarily unreliable indicators of productivity, but they were rather hit-and-miss compared with modern methods.

On either side of the Severn estuary, extending into the country east and west, a particular sort of dairy cow arose, broadly called the Severnside type, kept for cheese-making. Gloucestershire was right in the heart of this country. There, cheese-making has been concentrated in the vales of Berkeley and Gloucester since at least the thirteenth century and almost certainly much longer. The earls of Berkeley had a substantial dairy as early as the twelfth century. Cheese and butter markets were well established at Chipping Sodbury and Gloucester by the thirteenth. John Leland in 1535 said that Alney Island, just to the west of Gloucester, where King Edmund of Wessex and the Dane Cnut had met to divide up England in the summer of 1016, had 'goodly medow ground ... cheese there made is in great price'.

The old Gloucestershire cattle, whose milk made this cheese, were similar to their neighbours from Glamorganshire and closely related to the ancient dual-purpose Castlemartin type established along the coast of south Wales through Pembrokeshire, Carmarthenshire and south Cardiganshire. They all were distinguished by their white finching combined with a darker coat colour. The Gloucesters were mahogany, the Glamorgans slightly lighter and the Castlemartins (which became the modern Welsh Black and lost their dairying capacity) pure black.

Finching describes the contrasting white markings that certain breeds carry and is an ancient characteristic that takes different forms in different breeds. In the Gloucester it is a stripe along the back, under the belly and along the tail. The Gloucester's white stripe does not extend to the head, whereas the Hereford, native to the next-door county, has a white head, shoulders and dewlap, with a white switch to the tail, as we shall see in chapter 10.

Finching can run down the legs and flanks in a 'colour-sided' pattern, as for example in the English Longhorn and the Irish Moiled (hornless) cattle. In some breeds, colour-siding makes the cow look as if a white sheet has been thrown over its back and down the flanks, as in the now extinct Sheeted Somerset and the very much extant Belted Galloway. Colour-siding can also be seen in cattle whose body, ears, nose and feet are dark and the rest of the animal light-coloured. An extreme pale form of colour-siding is where the light colouring covers most of the body and the dark colour is restricted to the ears, nose and feet, leaving most of the animal white, as with the various types of white cattle, particularly the Chillingham (see Chapter 19) and the White Park.

By 1834, when William Youatt was writing, the Gloucester was in danger of being lost due to crossing with other breeds, first the Longhorn then the Shorthorn. Although he detected traces of the old breed in the first crosses, few native Gloucesters remained unadulterated. Crossing with Longhorns – particularly Robert Bakewell's* 'improved' variety – had become so common that although the cattle had gained some size and beefing qualities, it was at the expense of milk yield and quality of cheese. By 1900, the Gloucester seemed doomed to extinction.

It has been suggested that the feudal practice of heriot, by which upon the death of a tenant his heirs were required to surrender his 'best beast' to the lord, led to a concentration of good cattle in the hands of the landlords and hastened the demise of the Gloucester because ordinary farmers were forced to look to other breeds to stock their dairies. But that is more a swipe at the landlord class than a point with any substance. The beasts handed over to the landlord

---

* Robert Bakewell (1725–95), the famous agriculturalist and livestock breeder who farmed at Dishley Grange in Leicestershire and was at the forefront of the eighteenth-century revolution in selective breeding to create new types of domestic livestock for meat production. He was by no means the only innovator, but he was arguably the most skilful and influential and a masterly self-publicist; his incestuous breeding methods inspired, for good or ill, a host of imitators and have had a profound and lasting effect on livestock breeding.

represented a fine or fee that the tenant's heir had to pay to secure the continuation of the family's occupation of their land. In many cases the beast would be valued and the heriot paid in cash, or it would be put up for sale by the landlord and the tenant could buy it back. Landlords would receive numerous cattle as heriots and they would be unable to keep more than a few of them. There were so many cattle in the dairying parts of Gloucestershire in the sixteenth century that the vicar of Berkeley received up to 130 calves a year in tithes, notwithstanding that farmers who had fewer than six calves a year paid nothing. Bearing in mind that the size of a herd in the two hundred years between 1518 and 1713 was between nine and eighteen cows, and not every cow had a calf every year, there must have been hundreds of farmers keeping cows who did not have to part with any of them.

The high point of Gloucestershire cheese exporting was the middle of the eighteenth century, when the rich Gloucestershire pastures were supplying great quantities both locally to Bath and Bristol, and further afield to London and the colonies in West India, New England and the Caribbean. Double Gloucester travelled best and became rightly famous. It was made with the full-cream milk of two milkings when the cows were grazing summer pastures. The orange colouring was said to have been imparted by lady's bedstraw, which grew commonly in the ancient permanent pastures; it showed that the cheese contained high levels of beta-carotene from summer grass. But the colouring could only have been present in the milk for a small part of the year, and it would most likely have been pale yellow, not orange. In the sixteenth century, certain cheese-makers began to colour their cheese artificially by adding annatto, a vegetable colouring. This practice was then adopted by producers of Leicester, Cheddar and Cheshire, for without it their cheeses would have appeared a less attractive natural creamy colour. Thus they disguised the quality of their cheese by making it the same colour whatever the time of year, or what the cow had eaten.

True Double Gloucester was described in 1854 as having a 'blue coat, a golden hue on its edges, a smooth, close wax-like texture and a milk-rich flavour. It was not crumbly, did not part when toasted and softened without burning.' Apparently the best cheese of them all was Double Berkeley, found within a ten-mile radius of Berkeley: notably at Hardwicke, Haresfield, Leonard Stanley, Slimbridge, Stonehouse, Whitminster and Frocester. This disappeared in the last century until Charles Martell revived a version of it in 1984.

The less well-known Single Gloucester, or 'hay-cheese', was made from the winter milk of cows fed on hay, using whole milk from the morning's milking mixed with skimmed milk from the previous evening's. It tended to be eaten younger and closer to home, because it did not keep as well as Double, which was a bigger, firmer cheese, strong enough to bear the weight of the cheese factor – dealer – who would stand on each cheese; if any yielded, he would reject it as 'hoven' or 'blown'.

Even in the nineteenth century, the average Gloucester cow gave 500 gallons of milk in a lactation of around 300 days; this would make about 330 lb (150 kg) of cheese. But the cow's potential was barely tapped because it was expected to produce milk from whatever could be grown on the farm – grass from old permanent pastures, sainfoin or clover – and hay in winter. There was no fancy feeding and the herd was seldom, if ever, housed, except for some rudimentary shelter from the worst weather.

As an indication of what the Gloucester cow could have done under better conditions, Ladyswood Pansy 190, belonging to Lieutenant Colonel Henry Cecil Elwes of Colesbourne Park, 700 feet up on the Cotswolds, was recorded in the 1930s as giving 7,132 lb in her third lactation, 8,314 lb in her fourth and 9,214 lb in her fifth. Another cow, Colesbourne Bluebell 54, gave 6,000 lb in her first lactation, 8,712 lb in her second and 9,102 lb in her third, with 5 per cent butterfat and solids-not-fat of nearly 8.5 per cent.

Even compared with the yield of modern dairy cows, these figures are impressive. But when we take into account that the milking ability of all commercial dairy breeds has been increased by about half in the last 50 years by selective breeding and intensive feeding of bought-in protein, we can only speculate what the Gloucester could have done if she had been bred for yield. Although there is more to it than simple yield. The art of dairying is to balance the quantity and quality of the milk over a lactation with the cost of keeping the cow and breeding replacements, the longevity and health of the beast, ease of calving, and resistance to foot troubles. On any measure, if all these factors were weighed in the scale, the Gloucester would have come out pretty near the top.

But the Gloucester never had a chance to show what she was really made of because the breed was hit by two disasters from which it never recovered. The first was the terrible rinderpest epidemic of 1745–56. Cattle plagues were hardly unusual, but this was exceptional. The number of cattle that died or had to be slaughtered must have run into millions. In an attempt to stem the spread of the disease, the government ordered large-scale slaughter: Lincolnshire lost 100,000 cattle in 1746. Farmers were accused of flouting the slaughter policy and prolonging the disease, but it is hardly surprising when compensation was less than a third of the market value of the animal and it was far from clear that slaughter was having the desired effect. Many farmers preferred to risk their cattle catching the disease, rather than accept certain ruin by having them slaughtered.

The disease reached Gloucestershire in 1748 with an outbreak at Forthampton, near Tewkesbury. On orders from London, the county justices took drastic action: 80,000 cattle were slaughtered, and all fairs and markets for cattle in the county were closed. It is not clear how long the closures lasted, but fresh restrictions were imposed in 1752, and by the time the disease abated, the majority of the cattle in the county (and in the wider country around) had either died or been slaughtered. This hit the Gloucester particularly hard.

From an epidemiological perspective, rinderpest, which means 'cattle plague' in German, was an interesting disease. The fifth of the ten plagues of Egypt, it has flared up periodically throughout history, often at the same time as wars or civil upheaval – the 1745 outbreak accompanied the second Jacobite rising. In the 1890s, millions of cattle – 80 to 90 per cent of the entire population – died in southern Africa. The last major outbreak, in Africa between 1982 and 1984, prompted the United Nations Food and Agriculture Organization (FAO) to begin a worldwide eradication campaign.

Rinderpest killed almost all animals in a herd without natural immunity and was easily passed by contact, drinking infected water and sometimes on the air. It was almost always fatal. Cattle died slowly from fever, diarrhoea and liver necrosis within six to ten days of the first symptoms. The virus originated in Asia, and gradually between AD 1000 and 1100 transmuted itself into a zoonotic disease: a disease that originally only affects animals but develops into one that is transmitted to humans (and other species). In this case it became measles in humans and canine distemper in dogs. The virus itself is highly infectious while still alive but particularly fragile and quickly destroyed by heat and sunlight.

After the global eradication programme and the last case being reported in Kenya in 2001, the FAO declared in 2011 that rinderpest had been 'wiped off the face of the planet'. There have been no reported cases since then. It is only the second disease in history – the other being smallpox – to have been so eliminated. It remains to be seen whether the optimism of the scientists will be borne out, because the virus has a long history of mutation. Stocks are still maintained in specialized laboratories in 24 countries, much to the disquiet of the FAO, which in 2015 called for their destruction for fear of accidental or malicious escape, something that would destroy millions of cattle, particularly in America and western Europe, where animals have lost any natural immunity.

Although the Gloucester was hit particularly hard by the epidemic of 1745, the breed could well have recovered, given time, had there been the determination among farmers, and had it not been for the second contagion – Longhorn fever. The high prices for dairy produce caused by shortages following the rinderpest outbreak encouraged farmers to fill the vacuum with almost anything that would give milk. And after 1750, Gloucestershire farmers were too impatient to restock the empty dairies to revive their ancient breed. They simply bought cows from the nearest source, which at this time was the Midland counties. There were plenty of the old Longhorn dairy type to be had, as yet undamaged by the improvers, which gave adequate quantities of milk, high in butterfat for cheese-making, and would also satisfy the growing urban demand for beef by making a better carcase than the Gloucester.

The Gloucestershire dairymen were so eager to restock with these Longhorns that they ignored a truth William Marshall (1745–1818), the agricultural writer, early proponent of agricultural education and rival of Arthur Young (see p.189), pointed out in 1789: that 'it was the Gloucestershire breed which raised the Gloucestershire dairy to its greatest height' and that 'the breed had long been naturalized to the soil and situation; and certainly ought not to be supplanted, without some evident advantage, some clear gain, in the outset; nor even then, without mature deliberation; lest some unforeseen disadvantage should bring cause of repentance in the future'.

As so often happens in life, fashion also came into it. Anybody who aspired to be at the forefront of profitable farming had to have dual-purpose cattle. The specialist Gloucester, which for centuries had filled the county's milk pails, had made its reputation as a pre-eminent district for cheese and given the whey that fattened the Old Spot pigs in the county's orchards, suddenly looked old-fashioned. Progressive farmers could see the future and it did not include the Gloucester cow.

But they were mistaken. As time went on, the Longhorns coming out of the Midland counties had changed. They were not the same type they had been. What was increasingly emerging from those counties was stock that had had the milk bred out of it by Bakewell and his followers in their search for beef. This the Gloucestershire dairymen found too late, to their great cost. Professor David Low pointed out in 1842 that continual crossing over the previous half-century had turned 'a very large part of the cattle of Britain [into] a mixture of races, having no uniformity of character, and generally defective in some important points'. Then, as the nineteenth century progressed, the beef bubble burst and the Longhorn lost whatever appeal it had had, even to the grazier. It had already disappeared from the county of its creation; not a single Longhorn remained within a dozen miles of Bakewell's farm at Dishley, the place where the great man had once with such determination tried to create the perfect beef breed.

The Longhorn 'had acquired a delicacy of constitution, inconsistent with common management and keep; and it began slowly, but undeniably to deteriorate'. In other words, it had been so highly bred that it was unable to thrive under ordinary conditions of farming. 'It would seem as if some strange convulsion of nature, or some murderous pestilence, had suddenly swept away the whole of this valuable breed.' In fact, it had disappeared from most of its former strong-holds because in its 'improved' form it could not compete with the rising Shorthorn either for milk yield or as a grazing animal for beef. The improvers' work not only ruined the Longhorn; it also destroyed the hopes of a Gloucester revival. The Gloucester carried on in the hands of a few determined breeders, and demand continued for Gloucester bulls to cross with the Shorthorn to breed a high-yielding heifer, but its time seemed to have gone.

What looked like the final chapter in the Gloucester's story opened with the great London rinderpest epidemic of the 1860s, which killed about 80 per cent of the capital's dairy cattle. Nineteen years earlier, in 1841, the railway from Gloucester to London had opened, and such was the metropolitan demand for clean milk that a market in 'railway milk' for the capital opened soon afterwards. Cows could be milked in Gloucestershire in the evening and their produce be on London doorsteps next morning at a price that paid much better than cheese-making and without the labour. And in the Shorthorn, the dairymen of the West Country had found an animal well suited to the provision of liquid milk. Cheese-making had begun its long decline, to the point that the small amount of Double Gloucester that was still being made was sold as Cheddar.

As for the remaining Gloucesters, they would have gone the way of the Longhorn had it not been for the dukes of Beaufort at Badminton, who during the late nineteenth century gave leadership to a nucleus of breeders determined to preserve the breed. In 1909, the duke took some Gloucesters to the Royal Show, where they attracted wide interest, and by 1919 a society had been formed to promote the breed. It looked as if rehabilitation might be possible, until foot and mouth swept the county. It was said that the government's compulsory slaughter policy worked against the Gloucester, because the white finching made the cattle easy for the slaughtermen to see in the dark and allowed them to do their work well into the night. The final blow came with the Second World War, when most of Gloucestershire's ancient permanent pasture was ploughed up, apparently to feed the nation. By 1951, it was estimated that there were about 50 pure Gloucester cattle left, and their society had become the 'plaything of half a dozen landed gentry'.

By the time Charles Martell came onto the scene in the 1970s, the only remaining pure-bred herd belonged to the rather reclusive Misses Ella and Alex Dowdeswell of Wick Court, near Frampton on Severn. Their herd had survived in isolation on their farm, enfolded in a meander of the River Severn and run by the two women after the death of their brother, Eric, in April 1968. Charles Martell wrote to ask if they had any cows for sale and received a reply in less than a week saying he had just missed them: they had sold their 33-strong herd at a dispersal sale 'yesterday'. He was devastated. He had 'clutched at and missed something that was to be gone for ever'.

Despite this setback, he visited and charmed the Dowdeswell sisters and with their help traced all the buyers from their marked-up sale catalogue. It became clear that the cattle had been so widely dispersed that if the breeding of those few remaining animals was not coordinated, and records kept, the breed would almost certainly be lost. At a meeting of all those interested in 1973, it was resolved to re-form the society to try to save the Gloucester from extinction.

Joe Henson (father of Adam, of BBC *Countryfile* fame), who ran the Cotswold Farm Park and was a founder member of the Rare Breeds Survival Trust, had had experience of the breed when he was assistant manager at Earl Bathurst's farm. Bathurst's herd had grown to 170 cows, kept high up on the Cotswolds at Cold Aston. The only thing was that they were not all pure-bred, having boosted their milk yield by out-crossing. After the earl died in 1966, his herd was dispersed and Henson assumed the breed had become extinct, until three years later he read an article in *Farmers Weekly* about a bull called Gloucester, from the Dowdeswells' Wick Court herd, which was having semen taken for long-term storage as he was the last-known pure Gloucester bull.

Henson phoned Wick Court. In common with many country people of the time, the Dowdeswells didn't see the need for inside sanitation or running water in their ramshackle Elizabethan manor house, but they did value the telephone, which was answered by Alex. She would agree to nothing without the say-so of her older sister, Ella, and she told Henson to ring back.

Ella was forbidding: 'Our kid tells me you want to see our herd.' Henson persuaded her to allow him and his business partner to visit, with the clear understanding that she would not sell him any cattle. After sherry and biscuits in the house, served with old-world courtesy, the two visitors were taken to see the cows in the fields and gave a hand to get them in for milking. One was a newly calved heifer that had never been milked before, and, rather mischievously, the sisters asked Henson (who was a city boy, the child of two actors) if he would put the teat cups on her for the first time. 'I have never put on a cluster with greater care in my life; the heifer never moved. I had passed the test!'

After he had helped with the milking and turning the cows out to graze, the sisters 'got into a huddle' and emerged to announce that they would, after all, sell their visitors two cows – 'that one and that one' – because they seemed like 'nice young men'. Henson named the beasts Alex and Ella. They also allowed him to buy some semen from their bull, Gloucester, so he could inseminate the cows. Nine months later, Ella gave birth to a pair of twin bull calves, one alive, the other dead. But Alex proved to be barren. Ella had too much milk for one calf, so Henson got into the loose box to try to milk her out to relieve her discomfort. She nearly kicked him over the door and would never let anyone touch her udder thereafter.

A year later, he visited the sisters again, and while discussing their herd, Alex asked Ella to remind her which cows they had sold their nice young men. 'You remember? They had the barrener and the kicker.' When Alex broke her arm in 1972, the Dowdeswells reluctantly sold the herd and Joe Henson bought a cow called Nervous (renamed Alex) at the dispersal sale. These two cows were the foundation of his Bemborough herd using semen from Gloucester the bull.

Meanwhile, Charles Martell set himself to making Double Gloucester cheese from the milk of three Gloucester cows, which at first he milked by hand. He sold it wherever he could: local markets, farm shops and other outlets that were opening up to meet a tentative demand for locally produced natural food. He also revived the making of Single Gloucester, which had disappeared with industrialized milk processing, and obtained 'Protected Designation of Origin' status, which meant the cheese could only be made in Gloucestershire, ideally using the milk from Gloucester cows, although that was not essential because there were not enough of them to satisfy the demand.

Then, in 1994, Martell invented the cheese that has made his name. Dymock is in a part of Gloucestershire north of the Severn estuary, close to the Herefordshire border, whose climate is ideal for fruit growing, particularly apples, pears and plums. Martell set about finding, cataloguing and saving as many of these as he could trace. In the process, he came up with the idea of using the perry made from one old pear variety, called Stinking Bishop, to wash and flavour the rind of a new soft cheese he was developing and to which he gave the same name. It was a stroke of genius. And when it supplanted Wensleydale as Wallace and Gromit's cheese of choice, Charles Martell's fortune was made. He has gone on to invent another five cheeses – seven from cow's milk and one from ewe's milk – all connected in some way with Gloucestershire and its history. None is sold through the major supermarkets; all go through cheese wholesalers, smaller shops or direct to the public at markets or online. He has succeeded without having anything to do with big dairies. And Prince Charles has even awarded his cheese a richly deserved royal warrant.

Jonathan Crump is another, albeit younger, devotee who is keeping the Gloucester cow from oblivion. He started working at Wick Court after it had passed from the Dowdeswells' ownership and become one of the Farms for City Children started by Michael and Clare Morpurgo in 1976. He brought the Gloucesters back to Wick Court in 1992, nearly 20 years after the Dowdeswells' sale, and then moved to Standish Park Farm near Stonehouse, where he milks about 18 cows, making marvellous unpasteurized Double and Single Gloucester cheese. He is part of the Slow Food movement, which champions proper ways of rearing things, and his cheeses are authentic, made on the farm, and delicious.

His land is on a slope of the Cotswolds looking west towards south Wales and the Bristol Channel, the fields sweeping upwards from the farm buildings. This is ideal grass-growing country, with mild winters, overlying limestone on the edge of the Vale of Berkeley, the heartland of the old Gloucester. Jonathan is a natural stockman: quiet, even-tempered and gentle with his cows. He has never wanted to do anything other than work with animals, and it is immediately obvious when you see him milking in his cowshed. Nothing is rushed; the cows amble in for milking and Jonathan calls them by name, talking to them as they let down their milk.

This is not high-tech stuff. The byre and the milking equipment are basic and functional. The tank into which the milk goes is scrubbed stainless steel, clean and perfect for the job, but it's not fancy. The rooms where the cheeses mature on shelves are simple and have not cost a lot of money. Yet the cheese is wonderful and Jonathan has a market for everything he makes. If it was better advertised and promoted, there would be a waiting list and people beating a path to his door. Nonetheless, he is making a living from a few cows of a breed that was written off 50 years ago because it didn't perform as the intensifying industrial dairies required. Jonathan's championing of the breed is a heroic success and a rebuke to all those farmers, agricultural colleges and agri-businessmen whose perpetual cry is that farmers must 'get big or get out'.

It had been known for centuries that if you wanted a pretty wife with an unscarred face, you should marry a milkmaid. It had also been known that transferring a small amount of material from a smallpox sore to a scratch on the skin of another person would often give them immunity to the disease. This was called variolation, from *variola*, the Latin name for smallpox. But the procedure was dangerous because some recipients got a full-blown case of smallpox and died. Smallpox was a terrible worldwide scourge. It is hard for anyone brought up in the last half-century – a time when we have come to believe that illness and even death are preventable – to imagine living with a disease that affected roughly 60 per cent of the population and killed about 20 per cent of those who caught it. And those who survived carried disfiguring pockmarks for the rest of their lives.

But that was the way it was until Dr Edward Jenner (1749–1823), a surgeon living in Berkeley, and a few others – notably a Dorset farmer, Benjamin Jesty – showed that being infected by cowpox, which was much less dangerous, gave immunity to smallpox. Jesty experimented on his wife and children, something that was considered scandalous by his neighbours. They said he was inhuman, and mooed and threw things at him when they saw him in the street. People feared that introducing an animal disease into the human body would cause them to have bovine characteristics and grow horns or other appendages. Even though Jesty was proved right when his two older sons didn't catch smallpox after being exposed to it, it took some time for popular opinion to accept it.

James Gillray ridiculed the popular fears in a cartoon of 1802, *The Cow-Pock or the Wonderful Effects of the New Inoculation*, in which he showed people being rather brutally vaccinated with 'Vaccine-Pock hot from ye cow' and then growing various parts of cows from their bodies. In the background hangs a picture of the Golden Calf, implying that the vaccinators were in danger of worshipping their own creation rather than accepting nature as they had been given it.

On 14 May 1796, a Gloucester cow called Blossom, belonging to a Mr Dean of Berkeley, earned her place in medical history. Blossom had infected Sarah Nelmes, her milkmaid, with cowpox, and Jenner was sure that transferring pus from her blisters into a scratch on each arm of eight-year-old James Phipps, son of his gardener, would protect him from smallpox. Jenner came up with the name *Variolae vaccinae*, literally 'smallpox of the cow', to describe cowpox, and the procedure became known as vaccination. Phipps developed a fever, which soon passed, but had no other symptoms. Some time later, Jenner tried twice to infect him with smallpox by introducing infectious material into his system, but to no effect. He had tried the same thing with 16 other patients, so he was pretty confident it would work with Master Phipps. His vaccinations were described in his first paper to the Royal Society, in which he showed that all the people he treated gained immunity to smallpox. He went on to try it with a further 23 patients, all of whom acquired immunity. The medical establishment, as ever, was slow to accept his research and the Royal Society did not publish his initial paper, deliberating at length over his findings before they eventually accepted them.

In 1840, the British government banned variolation – using smallpox to induce immunity – and instead provided free-of-charge nationwide vaccination using cowpox. The various nineteenth-century vaccination acts from 1840 onwards consolidated the state's determination to eradicate the disease. It became compulsory for parents to have their infant children vaccinated before the age of three months. There was an undercurrent of opposition by a minority of parents who resented being forced by the authorities to submit their children to what was believed to be a potentially dangerous procedure. Safer vaccines, and more stringent compulsion, such as fines and imprisonment, reduced the number of refuseniks, but there was still some opposition, which bubbled away and boiled up from time to time – and still does today.

Jenner's discovery soon spread around Europe and the rest of the world. Even though France was at war with England,

Napoleon had his troops vaccinated; in return, at Jenner's request, he released English prisoners of war, remarking that he could not refuse anything to one of the greatest benefactors of mankind. In 1802, Parliament granted Jenner £10,000, and another £20,000 in 1807, after the Royal College of Physicians confirmed the widespread benefits of vaccination. The gift was worth £2.2m in today's money.

Blossom was not forgotten. She had her (rather amateurish) portrait painted by Jenner's great-nephew, and when she died, Jenner had her hide tanned and hung in his coach house. Reminiscent of Jeremy Bentham's stuffed skeleton in his box at University College, London, the hide now hangs on the wall of the St George's medical school library in Tooting, south London. Holy relics of the Jenner/Blossom cult continued in circulation long after the famous cow's death. At one time there were *five* of her horns in existence, and in 1896, hairs from her tail were put up for auction. Hardly surprising that a cult should have arisen over the eradication of the scourge of smallpox, which to most people in the world was nothing less than a miracle.

CHAPTER 3

The Shorthorn

George Bates was devoted heart and soul to farming. His highest ideal of happiness was the Horatian picture of a man owning and occupying a hundred acres, undisturbed by anything passing in the world outside. For those who regarded their estates merely as game preserves or props to their self-importance, and took no active interest in their cultivation, he entertained the bitterest contempt.

Cadwallader Bates, *Thomas Bates and the Kirklevington Shorthorns* (1897)

THE FARM-GATE WHOLESALE milk price in March 2018 in the UK was 30p a litre – little more than it had been eight years earlier. The cost of producing that litre depends on many things, but on average, with high-yielding cows, it is about 33p. Milk sells for between 70p and 90p a litre in retail shops – two to three times the price the farmer gets. So how do farmers milking cows carry on when they are losing so much money on every litre they produce, unless they turn their farm into a mega-dairy and hugely increase production?

The twenty-first of March 2017 was the first day of a two-day dispersal sale of the Brafell herd of Dairy Shorthorns at Penrith Farmers' and Kidd's auction mart. The cows to be sold were that half of the herd that had calved by the day of the sale. The remainder were to be sold on 23 May, when they too would have calved. The owner, John Teasdale, was 74 and had had two strokes; after the second, he had to be carried out of the milking parlour. His son was not interested in the slavery of milking cows, perhaps unwilling to make the sacrifice that had consumed his father, who had given his life, heart and soul to his herd of pedigree Shorthorns. For it is still upon people like John Teasdale that we depend to put milk, butter and cheese on our tables.

John's father, Thomas Teasdale, started the herd in the early 1930s at Uldale, near Caldbeck in Cumbria, high up on the northern edge of the Lake District, with glorious views of the Solway Firth to the north-west. John carried it on, latterly from a slightly better farm on the other side of the Solway at Kirkbean, near Dumfries. There is a glorious stubborn individualism in keeping Dairy Shorthorns in the first quarter of the twenty-first century. They are not the heaviest milkers, although they are better than they once were, and their milk is not as high in butterfat as some pure dairy breeds such as the Jersey. There is a similar prejudice against the farmers who keep them as there is against 'hobby farmers'. They are not commercial, they're a bit romantic; not real farmers at the cutting edge of modernity. In progressive quarters Shorthorns are sneered at as a recherché interest; any farmer doing the job properly wouldn't mess with the breed; instead he would have black-and-white Holsteins, *real* dairy cows that give proper quantities of milk. That's the thinking of the industrial dairy farmers. But it is not so certain that aggressively commercial farmers are right to dismiss a breed so deeply rooted in the history of British farming; that has been bred to make the best use of grass, especially on a hard farm.

A 'hard farm' means everything to a farmer, but it is not easy to explain. It can be anything from relatively infertile low-lying land where stock does not do as well as in other places, to fertile but high-lying soils where spring comes later and autumn closes in sooner. It can be a farm where the land is exposed to storms, or where snow tends to lie longer than elsewhere. It can be a farm facing north and east, rather than south and west, or on the wrong side of a valley, lying in shadow on winter days, white with frost where the sun cannot penetrate. It can be a farm with steep fields that are hard to cultivate. There can be myriad reasons why a farm is hard, but the essential thing to understand is that in more sensible times, when the value of a piece of land was related to what it would produce for human benefit, hard farms were cheaper than other farms and could give a skilful, ambitious young farmer a start in life.

Before the EU's single farm payment, inheritance tax benefits and a global asset price bubble drove the price of an acre of farmland into the realms of fantasy, a hard farm gave Thomas Teasdale a start in farming because he and his cows knew how to be resilient and thrifty, and if the farm couldn't produce it then they couldn't have it.

Every piece of land has a natural capacity to produce an annual increase derived from photosynthesis. Fluctuations in weather, human skill and effort can reduce or enhance the production; spending capital can have some effect. But many enthusiastic people have found to their cost that there is no limit to the amount of money that a hundred acres of English farmland will absorb. It can be drained, hedged, fenced, watered, fertilized and ploughed. All manner of fancy buildings can be erected on it, and expensive breeds of livestock can be bought to give more milk or breed more or better offspring.

But when it comes down to it, the only test of success is the amount of money left at the end of the year. And that depends, in almost every case, on the money the farmer doesn't spend. With dairy cows, the secret is to balance the milk they produce against the cost of getting it. That is where the Shorthorn justifies her existence. A good cow will produce about 5,000 litres in a 300-day lactation, against a Holstein, which gives on average about 7,000 litres. But the difference is that Shorthorn cows like Thomas Teasdale's will do that on home-grown grass and forage, with about 3 lb of cow cake a day, whereas the Holstein will need much more lavish feeding. In addition, a Dairy Shorthorn will last longer than a Holstein; she has smaller calves, so giving birth takes less out of her; and she is a better converter of roughage into milk, flesh and fat, so that at the end of her life her carcase will have a value, whereas the rangier Holstein will be little more than a bag of bones.

The Dairy Shorthorn went out of fashion when cake and fertilizer were cheap and milk was worth more in the wholesale liquid market irrespective of whether it was fit for something better than pasteurizing and homogenizing. Most herds simply went black and white. Many of the Shorthorn herds that remained had to consider other ways of surviving than selling to the MMB. The Teasdales got a small premium from a creamery to make industrial Cheddar cheese.

These Shorthorn herds contained good cows, well bred, with long pedigrees, but they were not worth much. At the

Teasdales' dispersal sale, the auctioneer made the point several times, trying to squeeze the last £20 out of buyers, that he was selling dairy cows for the same money 30 years ago: 'Come on, lads, get your hands in your pockets. These cows have come off a hard farm and can only improve. They just need a change of scene.' Pedigree livestock are still sold in guineas – one pound and one shilling (5p). The pound used to go to the seller and the shilling to the auctioneer for his commission. Now it is no more than an amusing anachronism, because the auctioneer wants more than 5 per cent commission.

It is usual not to milk dairy cows the night before a sale, so the buyers can see what their bags (udders) look like full of milk. At the fall of the hammer the auctioneer will usually offer the buyer the service of relieving the cow's discomfort by milking her before the journey to her new home. At Penrith, the auctioneer asked each buyer how much milk he wanted 'tekken off her'. The answer depended on how far the cow had to travel and how mean the buyer was. With some it was 'just a gallon', while others wanted her 'milking out'. One delightful roan cow, Brafell Lady 7th, went for 1,020 guineas; as she left the ring, the auctioneer shouted, 'How much do you want tekken off her?' and the buyer, quick as a flash, shouted back, 'About five hundred quid!'

Most of the Teasdales' herd was killed in the general slaughter during the 2001 foot and mouth epidemic, but the bloodlines were saved because some of their heifers were wintering away from the farm. They gave John Teasdale the nucleus of a resurrected herd. The average daily milk yield of all their cows is about 26 litres (6 gallons), with some giving over 30 litres (6½ gallons). The herd calves in the spring (spread over late January, February, March and early

April) to coincide with the growth of spring grass. The yield therefore peaks during the three months when the grass is at its most nutritious (late April, May, June and early July) and then naturally declines into autumn. The cows will mostly be dry by Christmas, in preparation for their two months' recuperation before the next calves are born and the cycle starts all over again.

The Shorthorn has occupied more attention and had more written about it than any other breed of cattle. Most of the highest prices paid for pedigree animals have been for Shorthorns, and the first bull to fetch 1,000 guineas was Comet, a Shorthorn sold by the Colling brothers in 1810. The Shorthorn has the oldest herd book, *Coates's Herd Book*, begun in 1822, and the oldest breed society, founded in 1872. Between the middle of the nineteenth century and the 1950s, Shorthorns were the most numerous breed of cattle in Britain, and are still the most widely distributed breed in the world. In the first cattle census, in 1908, 4.5 million of Britain's 7 million cattle were Shorthorns. And in the remarkable story of nineteenth-century pedigree cattle breeding, the Shorthorn is the principal character.

So where did this remarkable breed come from? For a partial answer, we have to look back three or four centuries into the east of England, Lincolnshire specifically, where a superior strain of milk cow first came to notice. One view is that the type emerged from the melange of localized varieties that had grown into regional types during the thousand years from the Roman recession to the fifteenth century. Another opinion is that during this long period, foreign stock was brought in to improve the native cattle, as was certainly known to be the case with horses. But there is no clear evidence as to the origin of what was described in various sources as the best English dairy cattle that could be found.

For many centuries, before and during the Roman occupation, cattle in the north-east of England seem to have been black. At the beginning of the third century, returning victorious from his Caledonian campaigns, the Emperor Septimus Severus arrived in York, where black beasts were to be offered up as a sacrifice for him in the temple of Bellona. He took it as a bad omen that animals of such colour were to be used, and forbade the sacrifice. But when the black cattle were freed by the priests, they followed him to the gates of his palace. The incident was seen (rightly, as it soon turned out) as presaging his death. Yet according to the Roman writer Varro, the Romans thought black was the best colour for cattle, followed by red, chestnut and white: 'for a white coat indicates weakness, as black indicates endurance'. Dun (or sometimes red) cattle were the commonest colour.

The dun cow is an ancient motif in English folklore, a fabulous animal that gives inexhaustible milk; hence the medieval saying that 'the dun cow's milk makes the prebend's wife go in silk'. It appears in the seventh-century Lindisfarne Gospels as a short-horned ox, and is associated with St Luke. It also comes into the hagiography of St Cuthbert. After wandering the north for some years with the saint's uncorrupted body, to fulfil their promise that they would never allow it to fall into the hands of pagans, the monks that comprised 'the fraternity of St Cuthbert' were searching for a place to bury him. They set down the bier carrying his coffin at a place (outside modern Durham) called Wurdelau – maybe Warden Law – and found they could not move it again. Aldhun, the Bishop of Lindisfarne, who was leading the group, decreed a three-day fast and prayers to determine the saint's will as to where his body should lie. Bede says that during the fast, St Cuthbert appeared in a vision to a monk named Eadmer and told him that his coffin should be buried at 'Dunholme'. But nobody knew where Dunholme was.

A little while later, the monks overheard a milkmaid asking another woman if she had seen her lost dun cow, and being told that the beast was grazing 'down at Dunholme'. They followed the girl and came upon a dun cow browsing in a wood high above a deep incised meander in the River Wear. This was where they buried St Cuthbert's body. Aldhun became the first Bishop of Durham, and founded a shrine and a monastery dedicated to the saint. The two women and the dun cow are commemorated in a sculpture built into the north-west end of the eastern transept of Durham cathedral – and in the names of numerous pubs in the area.

The story is more foundation myth than reliable account, because Dunholme must have been named *after* the cow was found there, not before. But it does tend to show that short-horned dun cattle were found in the north-east of England, particularly in County Durham, a thousand years before they came to prominence as the supreme dairy cow in the eighteenth and nineteenth centuries.

Robert Trow-Smith, never slow to find Dutch influence in almost all our improved cattle, argues that the origins of the Shorthorn were in north Holland or west Friesland. He refers to a passage in Robert Payne's *Brief Description of Ireland* (1589), in which Payne compares the best Irish cattle to the best he knew in England, which were 'the better sort of Lyncolnshire breed'. If they did come from the Continent, which is almost certain, it must have been either before or in defiance of the statute 18 Car. II, which in 1666 prohibited the 'importation of all great cattle' as 'a common nuisance'. This could explain why there is no official record of cattle being imported from Holland.

In 1707, in *The Whole Art of Husbandry*, John Mortimer described Lincolnshire cattle as 'pide': 'the best sort for the pail, only they are tender and need very good keeping, are the long legged short-horned cow of the Dutch breed which is to be had in some places of Lincolnshire, but mostly in Kent'. By the end of the eighteenth century, no trace of cattle of this type could be found in Kent, but they flourished in the north-east of England.

Any Dutch origin was long denied by elite Shorthorn breeders, who wanted to promote their stock as coming from 'some mythical, indigenous Old Adam of a beast ...' as Trow-Smith says. But it must be accepted that at some remote time, the Shorthorn came from the north-western seaboard of Europe, from Denmark to northern France, where for centuries a valuable type of cattle had existed. They were celebrated for the great quantities of milk they yielded, as well as an extraordinary aptitude to fatten, and were the basis of the large amounts of Dutch dairy produce exported in the late medieval period and into the modern era.

The cattle called Holderness, named after that spit of fertile reclaimed land between Flamborough Head in the north and Spurn Point in the south, were known for their size. William Lawson, the parish priest at Ormesby, in the North Riding of Yorkshire, explained in *A New Orchard and Garden* in 1618: 'The goodnesse of the soile in Howle, or Hollow-derness in Yorkshire ... is well-known to all that know the ... huge bulkes of their Cattell there.' These cattle were said to be coarser and heavier than the native breeds, capable of reaching enormous sizes if properly fed, but without the quality of carcase or early maturity of either the later Shorthorn or many of the native breeds. Their flesh was described as having a 'dark hue' or 'lyery' – not marbled – because the lean was not interlarded with fat.

George Culley explained in his *Observations on Live Stock* that 'lyery' or 'double-lyered', meant 'black-fleshed': 'for, notwithstanding one of these creatures will feed to a vast weight, and though fed ever so long, yet will not have one pound of fat about it, neither within nor without, and the flesh (for it does not deserve to be called beef) is as black and coarse-grained as horse-flesh'. He described such cattle as 'more like an ill-made black horse than an ox or a cow, bulky in the coarser points and small in the prime parts'. Their main merit was as extraordinary milkers, although their milk was lower in butterfat and solids-not-fat than the native Longhorn's. They were not good breeders either, their calves being described by Marshall as 'wide, square and scrawny across the hips' and what he called 'Dutch-arsed'. These big slow-feeding cattle produced the 'Lincolnshire Ox' exhibited at the University of Cambridge in the reign of Queen Anne. According to the advertisement, 'He was Nineteen Hands High and Four Yards Long from his Face to his Rump. The like Beast for Bigness was never seen in the World before. *Vivat Regina!*'

About eighty years later, a superior type, perhaps with origins in these coarse Holderness giants, emerged on the fertile banks of the River Tees. They were red, red and white, and (a favourite) strawberry roan. Their skin and flesh were 'fine and mellow' and their forequarters of great 'depth and capacity', according to Youatt. The best guess of their origin is that the coarse, leggy Holderness cattle had been taken in hand by a group of breeders in the Tees valley and, by unknown means, raised to be the astoundingly successful Shorthorn. The improvers almost certainly introduced a cross or two of other breeds into the Holderness to bring it closer to the ground, neaten its shape and improve its carcase fecundity and vigour. But, smaller though it was, the improved type was still capable of growing into a huge beast. A Mr Milbank of Barmingham in County Durham, one of the early improvers, slaughtered a home-bred ox at Barnard Castle in April 1789 at five years old. Dead weight, it was 177 stones 1½ pounds (almost 2,500 lb). The four quarters weighed 150 stones (2,100 lb), with 16 stones of tallow (224 lb), and even the hide weighed 10 stones 11 pounds (151 lb).

This began over two centuries of intense interest in the Shorthorn. Their pedigrees were recorded and studied with almost as much assiduity as those of the aristocracy. *Coates's Herd Book*, the Shorthorn breeders' vade mecum, and the tribes and families recorded in it, was taken almost as seriously as *Burke's Landed Gentry*. *Coates*, *The Book of Martyrs* and the Bible were the three essential reference books on many a nineteenth-century farmhouse table.

It is said that Bakewell's work showed that domestic animals' valuable characteristics were inherited rather than the result of feeding and management. But this may be giving too much credit to the great showman. Even though, at the time, pedigrees were recorded exclusively through the male line, it is hard to accept that intelligent breeders did not know the part inheritance from both parents played in the quality and performance of an animal. The Romans certainly recognized the value of inherited traits in both the male and female lines. Horace acclaimed the masculine: "'Tis of the brave and good alone/That good and brave men are the seed;/ The virtues, which their sires have shown,/Are found in steer and steed'; and Virgil the value of the maternal: 'The generous youth who studious of the prize,/The race of running coursers multiplies,/Or to the plough the sturdy bullock breeds,/May know that from the *dam* the worth of each proceeds.'

It might be that Bakewell showed that early maturity was an inherited trait, but it is hard to believe that cattle breeders could not see that for themselves before he based his reputation on it. It is true that much was done by eye and 'handle', and the ancestry of an animal was seldom known further back than a few decades in the memory of its breeder, but that is not to accept that breeders did not think it important to know how its ancestors had performed. They simply began to adopt a more rational scientific approach to measuring and recording the performance of their stock, in conformity with the spirit of the age.

Shorthorn pedigrees go back to 1737, further than any other breed of cattle, when the same Mr Milbank bought a red-and-white bull from the Aislabie herd at Studley, which rather unimaginatively he called Studley Bull. This animal is the ancestor of all the most celebrated Shorthorns. The earliest female pedigree starts in 1760 with a cow called Tripes, from which descends the celebrated Princess (formerly Bright Eyes) family. Her dam was bred in 1739 by a Mr Stephenson of Ketton and was probably a daughter of Studley Bull. The story of Tripes's descendants shows how much serendipity, coupled with the skill of the early breeders and an active market in cattle, went into making one of the most valuable and successful breeds the world has seen.

At Stephenson's sale in 1769, a John Hunter bought Tripes in calf. She produced a heifer, and because Hunter was 'a person in indigent circumstances [had no land]', he grazed his cow on the verges of the lanes around Hurworth, near Darlington, where he lived. In due time, the heifer was put to a bull owned by one George Snowdon and gave birth to a bull calf, which later became known as Hubback. Hunter sold the mother and her calf at Darlington market to 'a Quaker', who sold them on to a Mr Basnett, a timber merchant. On his way home, Basnett resold the calf for a guinea to a blacksmith named Natrass at Harrowgate near Darlington. Basnett took the cow home, but grazing on unaccustomed good land, she quickly got too fat to breed. This 'quick feeding' characteristic (using her food efficiently) she had passed on to her son. Natrass gave the bull calf to a young man at Hornby who became his son-in-law. The animal was much admired 'by all who saw him running in the lanes'. He was then bought by a William Fawcett of Haughton Hill.

The next part of the story has passed into myth and been retold in various versions. Youatt's version, which he recounts in *Cattle* in 1834, was, he says, told to him by Mr Robert Waistell himself; it had appeared in a similar form in Bailey's *General Survey of the Agriculture of Durham* in 1810. Another version is to be found in Cadwallader Bates's book[*] about his grandfather, recounted nearly a hundred years later without attribution, although he did have all Thomas Bates's letters and papers.

Mr Waistell used to ride almost daily by the meadow where the young bull was grazing, and so admired the beast that

---

[*] *Thomas Bates and the Kirklevington Shorthorns.*

after a time he offered to buy him from the owner. The price Fawcett asked of eight guineas seemed excessive to Waistell, who refused to pay. However, as the weeks went by, Waistell grew fonder and fonder of the animal, until one day he took his friend and neighbour Robert Colling (1749–1820) to have a look at him. Colling rather grudgingly acknowledged that the beast had some good points, but there was something in his manner that caused Waistell to suspect he thought more highly of the beast than he was prepared to admit. So next morning Waistell hurried round to the bull's owner, made the bargain and paid the money. Hardly had he done the deal when Colling turned up, and was disappointed to find he had missed his opportunity. On the way home, however, he persuaded Waistell that they should go 'halvers' on the bull, and Waistell sold him a half-share for four guineas.

Some months passed, and either Waistell's admiration for the animal waned or Colling had worked on him, because the partners sold the little bull to Robert's brother Charles (1750–1836), who had recognized in the animal just the qualities he sought for furthering his breeding programme. Once he had him in his possession, Colling refused to allow anyone else to use the bull. To Waistell's chagrin, Colling asked a fee of five guineas to serve one of Waistell's cows. Waistell refused to pay the fee and reminded Colling that he had sold his half-share for four guineas. But he was no match for the formidable Colling brothers.

Fate got its own back on Colling, however, because in 1787, 'by an extraordinary error of judgement', he sold the bull when it was ten years old to a 'Mr Hubback from Newbiggin in Northumberland' for 30 guineas. (It is likely that the purchaser was a Mr Huggup of Spital House, North

Seaton, where his surname was pronounced locally Hubback.) The animal lived on to serve cows in the district for another three or four years; it is from this bull that almost all the superior Shorthorns descend. In 1790, the year before the vigorous beast's death, Thomas Bates, whose influence on the Shorthorn surpassed all others in the nineteenth century, saw him and marvelled at the evidence of his remarkable prepotency still evident in his calves, even those that were bred out of unremarkable cows.

The Collings' breeding led to another famous bull, Foljambe (grandson of Hubback), who was the father of both the sire and dam of Favourite, which, when put to a 'common cow', engendered the celebrated Durham Ox born in early 1796. By February 1801, at five years old, this animal weighed 216 stones – 1.35 imperial tons. Deadweight, the carcase would have been 168 stones. The unfortunate animal was sold for £140 'into a fate of itinerant exhibition' to a Mr Bulmer of Harmby near Bedale, who had a special carriage made to convey it around for showing.

After five weeks, on 14 May 1801, Bulmer sold the bull and its carriage for £250 to a Mr John Day, who on the same day was offered and refused 500 guineas for the ensemble. A month later, Day refused £1,000, and on 8 July £2,000, preferring to travel with his bull around 'the principal parts of England and Scotland', which he did for six years. At ten years old, the ox weighed over 270 stones live weight (1¾ tons) – an estimated 220 stones deadweight. On 19 February 1807, the poor creature dislocated its hip and, after being jolted around for a further two months, had to be slaughtered on 15 April. Despite the considerable pain he must have suffered during his last two months, his carcase still weighed

186 stones (1.16 tons), of which the tallow was 11 stones (154 lb) and the hide 10 stones (140 lb).

The poor old Durham Ox was not the only huge Shorthorn beast to be taken round in a cart and shown off. Even Robert Colling himself joined the fashion and traipsed the countryside with 'The White Heifer That Travelled'. She was a freemartin, which is the female of male and female twins that has been made sterile in the womb by the blood vessels of the foetal sacs joining together and causing the foetuses to share the same blood circulation. Where both are of the same sex this does no harm, but if they are male and female, the male hormones give the female masculine characteristics. Ninety per cent of female twins are affected and born infertile, whereas the male twin is usually unaffected, although sometimes his testicles might be slightly smaller. Freemartins behave rather like bullocks (steers), being neither wholly male nor wholly female.*

The most significant improvement in the breed came about when the Colling brothers got to work. Around 1780, Charles started collecting good females of the type he wanted. The Collings were not the first to see the eighteenth century's need for 'economical flesh' – they had absorbed the gospel at the shrine of their old master Bakewell – but they were the first to make a success of the Dishley incestuous breeding methods, which they applied to their Shorthorn stock. Bakewell never achieved the results he had hoped for from his efforts with the Longhorn (or the Leicester sheep,

---

* Freemartins appear in Aldous Huxley's *Brave New World* as women who have been deliberately sterilized in the womb by the administration of hormones, as part of the government policy that requires 70 per cent of the female population to be freemartins.

for that matter), whereas the Collings' transformation of the local cattle into the Shorthorn was a spectacular and lasting success.

Purists at the time, and since, tried to denigrate their work by doubting the bull Hubback's purity. But as Charles Colling always admitted that he had used a cross of at least one other breed later in his career, complaints that the breed was somehow less valuable because its pedigree could not be traced back unalloyed to the Ark would condemn every superior breed of domestic livestock that has ever existed.

Unlike the nineteenth-century improvers, who have bequeathed us an embarrassment of riches of genealogical material, those from the eighteenth century left us practically nothing. It is probably trite to say that the story of his art is to be found in an artist's work, and had he wanted to express himself in a different way, he would have done so. It was no different with these early cattle improvers. They were single-minded, practical men of vision who brought all their talent and energy to their work in an atmosphere of intense competition in the early heady days of the Agricultural Revolution. And for those who got it right, there were ample fortunes to be made. So they were hardly likely to risk revealing their methods and allow their competitors to steal a march. If the Collings had learned nothing else from Bakewell, they certainly knew the benefit of keeping their methods obscure.

At Charles Colling's sale on 11 October 1810, 47 cattle (including bull and heifer calves under a year old) were sold for £7,115 17s. His light roan bull Comet, a son of Favourite, was the first bull ever to fetch 1,000 guineas – paid by a syndicate of four breeders. Eight years later, at his brother Robert's sale, 61 head of cattle made 7,484 guineas. By this time, prices were falling, as the agricultural distress brought on by the ending of the war with France began to bite. In the two years from 1814 to 1816, farming passed from prosperity to an extreme depression that did not abate until the accession of Queen Victoria in 1837.

The Collings, like all cattle breeders, had to choose between milk and beef. They chose beef. Thus began a division of the Shorthorn into two distinct types. Later pedigree breeders such as John Booth of Killerby and Warleby, and noblemen such as Lord Althorp and Sir Henry Vane Tempest, furthered the divergence. By 1817, animals taken to Scotland from Booth's herd became a separate type, the Beef Shorthorn, and it was left to Thomas Bates of Kirklevington, in County Durham, to maintain the milking qualities of the breed that became the Dairy Shorthorn. This gave rise to the saying in Shorthorn circles, 'Booth for the butcher, Bates for the pail'.

Although Bates recovered some of the milking qualities of the old coarse Holderness cattle that the Collings' breeding had damaged, his famous pedigree families – Princess and Duchess being the most notable – never fully matched their ancestors' yields in the dairy, although their milk was of better quality. He was criticized for being more concerned with style over substance, and he never overcame his pedigree cows' shortcomings in the pail. It was left to the commercial breeders to retain some of the early Shorthorn's exceptional milking qualities.

One commercial Cheshire dairyman was reported as saying at the Newcastle Royal Show of 1864 that 'if he wanted milk he would rather have his stock related to the Cheshire cows of 1800, to the Ayrshire, or even the Welsh cow … than to Royal Dukes and Duchesses, and would prefer their being matched to the son of his neighbour's best milking cow than to a bull of Bates's or of Booth's'. H. H. Dixon remarked in the middle years of the nineteenth century* that any allusion to good milking pedigree in a sale catalogue of Shorthorns was regarded as an apology for doubtful or unfashionable blood. Such was the pedigree breeders' disregard for the milking qualities of their stock.

The fact was that the Shorthorn breeders came up against the old truth that not even Bakewell could overcome: you cannot have a cow that is equally good at producing milk and meat. Had they been able to achieve this, the British Friesian would have found it hard to oust the Shorthorn from the dairies of Britain after the Second World War.

* H. H. Dixon, 'Rise and Progress of Shorthorns', *Journal of the Royal Agricultural Society of England*, i/ii, 317.

CHAPTER 4

The London Dairies

BEFORE REFRIGERATION AND easy transport, in most big cities, even in the grandest parts, people would have had milk cows living close by. Even if they didn't come into daily contact with them, they would have been able to smell them. At one time there were dozens of these city dairies, where cows were kept and milked on the premises and their milk either delivered house to house or sold from a shop attached to the cowshed. In smaller cities and towns the cows were kept in the countryside around, close enough for the milk to be brought in daily.

In London, there was a tradition of Welsh country people, especially from Cardiganshire, coming to the capital to make their fortune in the milk business. By 1900, over half London's dairies were owned by Welsh emigrants, who spoke Welsh among themselves, kept their chapels and held fast to the Land of their Fathers. Once they had made enough money, many of them returned home and set themselves up on small farms.

These dairies had long been a feature of the metropolis. In 1694, a Mr Harrard, who lived at Baumes in Hoxton, bought up newly calved cows from the countryside around London, milked them through their lactations, and when they were dry, sold them fat to London butchers. He kept 300, sometimes 400 cows. The London cow population grew with the city. By 1794, there were estimated to be 8,500, nearly half of which were kept between Paddington and Gray's Inn Road, with about 2,500 in the East End. William Marshall reported in 1798 that one dairyman kept 1,000 cows in milk, while there were three or four with 500 each. While the trade seems to have been dominated by a few large operators, it was always open to small producers to try their hand as well.

William Youatt wrote about the London dairies in 1834. After travelling the kingdom gathering information for his *Cattle* book and being welcomed wherever he went, he was disappointed to find that the gates of the 'overgrown milk establishments' of the capital were closed to him. Assuming they had something to hide, he determined to discover what it was. By this time the metropolitan herd ran to about 12,000 cows, almost all of the Dairy Shorthorn breed. They supplied Londoners with 'new milk', a product that sounded wholesome and healthful, but that hardly ever made it unadulterated to the breakfast tables of the populace.

Youatt noted the common practice of most, if not all, of the 'little dairymen' who kept half a dozen cows, of putting by the evening's milk until next morning, skimming off the cream, adding a little warm water and selling the collation to the public as that morning's milk. The real morning's milk was also put by, skimmed and warmed a little before being sold as the evening's milk. Retail milk sellers, who didn't keep any cows of their own, bought whole milk from the large dairies, but whatever they did with it before they sold it, Youatt was sure it was not to the benefit of their customers.

Now, of course, people have been conditioned to accept, even prefer, skimmed milk, and would make no complaint. To be fair to the dairies, they had to be skilful in meeting fluctuating demand for their highly perishable product. If it didn't sell quickly, they only had a short time to turn it into cream or butter.

Most of the smaller and medium-sized dairymen bought newly calved cows because they did not have the facilities to breed from them. Once a cow's daily yield fell below four quarts (a gallon), she was dried off and fattened for sale to the butcher. They could not afford to carry passengers. That was why the Dairy Shorthorn was in favour, because she was, without doubt, the best dual-purpose cow then available. Her milk was high in solids, although not as high as the old Longhorn she replaced, and she would readily fatten on clover hay and linseed cake once she stopped milking. This one-lactation system was wasteful of good milking cows and added considerable expense for the dairymen. Newly calved cows from the dealers cost about £20 each and were highly variable milkers. The buyer could not tell from looking at them how well they would milk: the size of a cow's udder is a poor guide to whether or not she will be a 'deep milker'.

The larger producers, with the space and facilities to breed their own replacement cows, had an advantage over their smaller competitors. They knew which animals were worth keeping because they had milked them through many lactations – often up to seven or eight – a much better practice than selling good milking cows and replacing them with animals of unknown quality.

They also had the advantage that the dealers who supplied most of the cows to the London dairies preferred to do business with larger enterprises because they could move cows in larger lots, often at a better price than the poorer dairyman could pay. A few of the biggest dairies had buildings that could accommodate five or six thousand cows, and the dealers used them as staging posts for their animals until they could be sold on. The dairy owner would charge a shilling a night for a cow's accommodation, have his milkmaids milk the cows and keep a record of their yields. He knew better than anyone else – even the dealer – which cows were yielding well, had a good temperament and didn't kick. A cow gives her best yield and quality of milk during her third or fourth lactation, and it was these that fetched the best prices.

Youatt got his own back on the London cow-keepers who had snubbed him by showing, with calculations, that they were defrauding their customers. He estimated the average yield across all the dairy cows in London to be 9 quarts (2¼ gallons) a day. Multiplied by 12,000 cows made it 108,000 quarts, and multiplied by 365 days that came to 39,420,000 quarts a year (9,855,000 gallons). Milk was sold at 6d. (2½p) a quart, which made the annual value of all milk sold in London (without the cream and butter) £985,500 a year. If this was divided by the 12,000 cows, he arrived at the 'strange and incredible sum of more than £82' output from each cow per year. This, said Youatt, proved 'the rascality that pervades some of the departments of the concern'. He acquitted 'the wholesale dealers of any share in the roguery' and laid the blame squarely on the retail dealers.

The conditions in which the cows were kept were remark-able for the time. An article on 'London Dairies' in the *British Farmer's Magazine* of February 1831 describes the large one run by a Mr Rhodes, 'farmer, near Islington'. The cowsheds faced east on a piece of gently sloping ground of about two or three acres in extent; they were light and airy, with panes of glass above ventilation shutters set on solid walls. Each shed held four to five hundred cows and ran up and down the slope to allow gutters to carry away their waste. The cows' drinking water also ran downhill from trough to trough along the whole length of the shed. Piped water was rare at the time because lead piping was expensive, and letting it flow freely avoided the labour of carrying water to the cows or having to let them out to drink twice a day from water troughs.

There was a separate range of buildings around a square yard where the cows were fattened after they had finished giving milk, and further sheds to accommodate the pigs kept to consume surplus skim milk, which was held in a huge tank twelve feet deep and six feet in diameter. The milk quickly went sour, but it didn't matter because that was believed to be more nourishing, although it is not recorded what the smell was like. There were pits for storing roots, straw, hay and grains, and a place for cutting straw into chaff. At the lowest part of the slope was the deep pit into which all the dung was emptied. There were cart sheds, stables and other buildings; in fact everything that was needed to run an efficient, almost industrial enterprise, remarkably modern for its time. Nowadays it would be described as 'zero grazing', where everything the cows ate was carried to them.

The unfortunate cows were tied in their stalls during the whole time they were in milk. Some of them had been chained there for more than two years. The only exercise they got was standing up and lying down. Youatt was critical of their treatment and compared it to a more enlightened dairy owned by a Mr Laycock, whose cows were let out twice a day for up to three hours, to drink and exercise; during the growing season, they grazed from six in the morning until midday and from two o'clock in the afternoon until three o'clock the next morning. Rhodes claimed his cows gave more milk, but Youatt associated being tied in a stall for 24 months with disease, foot and digestive troubles and a deterioration in the quality of the milk. Without going so far as to call it cruel, he did not have much good to say for such an unnatural system of cow-keeping, which attracted the same criticism as is raised against modern indoor systems of dairying.

The main foodstuff for metropolitan cows was the abundant and cheap supply of spent brewers' and distillers' grains, which were available all year round, although cheaper in spring and autumn when the dairymen stocked up on them. Brewers' grains are the germinated grains of cereal (usually barley in Britain) left over from brewing beer. They have been soaked in water, allowed to germinate and then dried to produce malt. The malted grains are then milled and steeped in hot water (mashed) to transform the starch into sugars. The resulting wort is boiled, filtered and fermented to produce beer. The residue, after the sugars have been extracted from the grain, is concentrated protein and fibre and suited to the feeding of ruminants. Not all grains have the same feeding value; it depends how much goodness has been extracted in the brewing process. As Youatt remarked, 'the dairyman must know his brewer'.

Brewers' grains have been a valuable part of the feeding of livestock for thousands of years. Large estates, farms and monasteries had them left over from their brewing. But with the growth of the cities and towns during the Industrial Revolution, huge quantities of grains were being produced that brewers had to dispose of. Town dairies were often attached to a brewery, and the still-warm grains could be ensiled in brick-lined watertight pits, trodden down to expel the air and sealed with a covering of eight or nine inches of soil to keep them moist in summer and protected from the elements in winter. Warm grains would continue fermenting until, having used up the available oxygen, they would be pickled, like silage, and preserved until the pit was broken open. Fermentation improved their feeding quality, and when properly stored, they would keep for two years or more. Sometimes salt was added to improve the palatability and provide minerals for the cows.

A bushel of grains a day (about 35 lb), spaced out in five feeds over twenty-four hours, would stimulate milk production. This was combined with 10–12 lb of chopped hay and two bushels of mangolds, turnips or potatoes, depending on the time of year. Another commonly used high-protein feeding stuff was linseed cake, the residue from pressing linseed (flax) to obtain the oil, although it was more useful for fattening the cows at the end of their lactation. It had the property of making their coats shine.

There was a clever division of labour in the dairies. The dairymen looked after the cows and made them available to be milked by the milk dealers from about four o'clock in the morning until six thirty in the evening. The retail milk dealers would order the milk they needed for that day's delivery and the dairyman would allocate to the dealer's milkers the number of cows he thought would supply that amount. Whatever milk they took was carefully measured in the measuring room, and if the yield was more than the retailer had bargained for, the surplus was kept by the dairyman to make up any deficiencies in other retailers' yields. If any was left and could not be sold that day, it was skimmed and its cream made into butter, with the whey being fed to the dairyman's pigs.

There were dairies in a few other major cities in the kingdom, with a 'leviathan dairy' in Glasgow owned by William Harley, a Scottish entrepreneur. Harley sold water from springs rising on his small estate, and ran public baths. But his dairy enterprise – Willowbank Dairy, at the top of West Nile Street – was without doubt his most impressive undertaking. His byres, wrote H. H. Dixon, 'lay among a frowning forest of chimneys, and [were] reached through mud and mire, now over a tram-road, now across a canal, and finally past a manufactory where horses [were] boiled into glue by the score'. There was a viewing gallery to which 'princes, noblemen and gentlemen from almost every quarter of the globe' came, paying a shilling a ticket to admire the lines of cows being fed four times a day from feed-trolleys. Grand Duke Nicholas of Russia (later the tsar), and Archdukes John and Lewis of Austria were among the distinguished visitors. Harley was a masterly self-publicist; he even wrote a book in 1829 celebrating his enterprise, *The Harleian Dairy System*.

He kept over a thousand cows in sheds holding about a hundred each, managed by two men to a shed, feeding and mucking out and grooming the animals daily. He gave his sheds names: The Parlour, The Thistle, The Holloween (*sic*), The Waterloo, The Malakoff and so on. At the peak of his enterprise, in 1860, before the European rinderpest epidemic struck, he had 1,700 cows in milk. He fed generously for yield on the sound Scots principle that a 'coo milks by the mou" – you won't get out what you don't put in – with draff (the Scottish word for distiller's grains), steamed turnips, cut hay, straw, ground maize meal and 6–10 gallons a day each (as much as they would take) of pot ale, the high-protein liquid left after the first distillation. In summer he cut grass and brought it in from his meadows.

He devised an ingenious labour-saving arrangement for clearing away 'the scourings' – cow muck and urine. Each cow was measured when she arrived in the byre and allocated a stall that exactly matched her length so that her muck and urine would fall away from her into gutters that ran behind the close-packed rows of cattle. These were regularly flushed into a tank, the contents of which (what we would now call slurry) was discharged along miles of pipe out of the city to fertilize the meadows where the grass grew.

Each dairymaid was allotted 13 cows in full milk (more if the cows were giving less). The high yielders were milked three times a day and the professional dairymaids were encouraged to sing to their charges because Harley knew 'cows are partial to a pleasing sound'. The average yield of all cows and heifers in milk was one and a half to two gallons a day. Every Friday the yield of each was recorded and the quality tested by 'lactometer'. Seven milkers lived on the premises and the rest came in from the city at milking times: five o'clock in the morning, noon, and six o'clock in the evening. To milk so many cows would have needed more than 70 people just for the milking, and another 25 or so to look after the cows, not to mention the men employed in the fields.

The coming of the railways and the devastating outbreak of rinderpest in 1860, which carried off 80 per cent of the cows, damaged the city cow-keepers' trade, but it did not destroy it. It would have recovered had it not been for motor transport, refrigeration, and the increasingly stringent public health measures designed to eradicate childhood tubercu-losis. Some dairymen lasted until the Second World War, delivering warm milk twice a day, with the attraction that it came from cows kept on their own premises.

Even as late as 1950, there were more than 700 dairies selling fresh milk in London. Jones Bros in Middlesex Street in the East End is the last of the Welsh dairies still oper-ating in the city. The business, which was started in 1877 by Henry Jones, has survived two wars and being undercut by supermarkets and convenience stores. The current Henry Jones, with his sister Lucy, is the fourth generation, albeit milk is now only a part of their operation. They have cleverly expanded into all kinds of fresh food and groceries delivered to homes and offices across London, from Canary Wharf to Kensington. They have the edge on the supermarkets because their delivery is free and no order is too small. It's a remark-able and inspiring story of survival against the trend by giving a service to the public.

Not all the milkmen have disappeared from London. The enduring appeal of having fresh milk delivered daily to your door, guaranteed quality, and for older people, having someone call at their house every day has ensured their sur-vival. Who knows, in these days of home deliveries of almost everything, the town milkman might even make a comeback.

Although milking cows on the premises has disappeared, dairying in the metropolis has metamorphosed into a new way of providing Londoners with the goodness that is to be had from cow's milk. From a very low point after the Second World War, when it looked as if we would go the way of the US into industrial, pasteurized, denatured cheese, there has been a renaissance in cheese- and butter-making in and around London. This owes a great deal to Randolph Hodgson, who started Neal's Yard Dairy in the late 1970s to promote and revive what was left of Britain's farmhouse cheeses. His powerful guiding spirit has inspired a loosely-connected web of artisan cheese-makers championing raw-milk cheese in a dogged and clever campaign for over 35 years against the commercial power of the big dairies and a Food Standards Agency determined to force all milk and cheese to be pasteurized. His support for small, high-quality cheese-makers has led to a revival of the demand for regional cheese and provided an alternative to the dispiritingly uniform products of industrial dairy processing. His shops in London and his wider influence all over the country have shown what proper cheese should be.

One of the alumni of Hodgson's academy at Neal's Yard is William Oglethorpe. He makes cheese, yoghurt and butter at Spa Terminus under the railway arches beneath the main line out of London Bridge station, and sells it at Borough Market from his shop called Kappacasein. Oglethorpe has been an urban cheese-maker since he founded the dairy in 2008, and now employs 12 people, inspired by the alchemy of cheese-making: the magical transformation of perishable milk into durable cheese. Twice a week he leaves London at 4.30 a.m. to drive to Bore Place at Chiddingstone in Kent, where he collects 600 litres of raw milk, fresh from the morning's milking.

To save time, he cultures the milk in the churns before it starts the journey back and aims to have the still-warm milk ready to go into the copper cheese vat by the time he arrives. He produces a number of 'un-tampered with' (as he puts it) cheeses every week: nine 6 kg wheels of Bermondsey Hard

Pressed, a bit like Gruyère, which is matured for six to twelve months, washed and turned twice a week; Bermondsey Frier, made to be fried in 100 g slices and a copy of an Italian Formaggio Cotto recipe that browns on the outside and remains 'squeaky' clean on the inside; and a ricotta, made by heating the whey left over after making the other two cheeses. He also makes natural yoghurt using his own starter cultures.

Another alumnus of Neal's Yard is Blackwood's Cheese Company, which is an offshoot of the Commonworks farm at Bore Place. They make, among others, a 'convict series' of cheeses named after English malefactors who were transported to Australia in the nineteenth century for stealing cheese. Edmund Tew is a small lactic cow's milk cheese named after a man who pleaded guilty to stealing a loaf of bread, some cheese and beer from the dwelling house of John Boot at Leicester. He was transported for seven years in 1829. Their other convict cheese is William Heaps, a fresh lactic cow's cheese whose namesake was sentenced in 1838 at Lancaster Quarter Sessions, also to be transported for seven years.

It is unlikely that there would have been the huge increase in the production and consumption of fresh milk without the revolutionary invention of the mechanical milking machine. Few innovations have had such an effect on farming. Before this, each cow had to be milked by hand, a laborious business that limited the number of cows that could be kept on one farm, unless sufficient people could be found to milk them. It was usually women who did it; they had a gentler touch and nimbler hands than men, and as a result, milking came to be seen as within the wife's domain, part of the household management. Hand-milking is a pinch-and-squeeze operation that is quite difficult to master. The teat is held in the palm of the hand with the forefinger and thumb making a tight circle round the top of it, just below the udder. This closes off the top and stops the milk from being forced back into the udder. Immediately after the top is pinched, the three lower fingers squeeze the teat to expel the milk. Two teats are milked at a time, each hand alternately pinching and squeezing in a rhythm.

It is important to strip out properly all the milk from a cow's udder at each milking. This encourages her to produce milk and discourages disease. So it became the practice, rather like moving seats at a dinner party to reanimate the conversation, for hand-milkers to move their stools around the shed after they had milked their own batch of cows, and try to milk out the 'strippings' from their neighbours' animals; sometimes bonuses were paid to those who could milk out the most.

But as yields increased, each cow took longer to milk; even an experienced milker could manage no more than five or six cows in full milk in an hour, having to rest their hands every half an hour or so and sometimes hold them against a cold stone to relieve the pain from strained sinews. There was a crying need for a serviceable, reliable machine to liberate the cow-keeper and milkmaid from twice- (or even thrice-) daily drudgery. During the middle years of the nineteenth century, many inventors tried their hand at making a workable machine. Scores of patents were issued for contraptions of dubious utility. There were those that mimicked the pressure applied in hand-milking and used various types of catheter that had to be pushed up the teat past the sphincter that sealed it from the udder; others mimicked the calf's sucking by creating a vacuum at the end of the teat.

Many dairymen would have nothing to do with the insertion of tubes into the teats, claiming, with some justification, that the practice introduced disease, damaged the udder and was painful for the cow. On the other hand, there was much reasonable objection to the sucking machines because they were thought to deform the cow's udder and also cause disease. The main problem with sucking was that it used a continuous vacuum, which bruised the tissues in the udder and inflamed the teats by drawing blood into them. One ingenious but unwieldy invention was a large latex bag that fitted over the whole udder and, by creating a vacuum, drew milk from all four teats at the same time. This was unsatisfactory because it tended to spread disease from one quarter of the udder to the others, took no account of the varying amounts of milk in each quarter, was hard to keep clean and was painful for the cow.

Some progress was made with the invention of an assortment of machines that had individual latex teat cups for use with a vacuum milker. But they all had the fatal defect of using a continuous vacuum, which applied constant pressure to the teats, damaged their tissues and tended to cause disease.

Eventually, in 1895, Alexander Shields of Glasgow solved the problem when he invented the pulsator, a device that regularly broke and remade the vacuum. Its pulsing mimicked the natural squeezing and releasing involved in hand-milking, and caused the milk to flow intermittently from the teat while massaging it at the same time. He installed this in a Thistle Mechanical Milking Machine, powered by a steam-driven vacuum pump. Proving how silly experts can be, when the Thistle machine was demonstrated at the Hamburg Exposition in 1898, it was reviewed by a Dr Benno Martiny, a self-important dairy scientist of the time, who dismissed the device because, as he observed through the glass inspection tube inserted in the rubber pipe, it caused the milk to flow intermittently. He failed to see that the pulsator was the single most important invention in dairying, which led to a properly workable milking machine.

The first machines collected the milk in 'units', portable stainless-steel buckets with a sealed lid like a Kilner jar, to which the pulsator and a cluster of four teat cups were attached. The milker connected a rubber tube to a tap on a metal pipe, which provided the vacuum and gave the suction. When the unit was full, it had to be carried into the dairy and the milk poured through a cooler, which was a kind of radiator with metal ribs, with cold water running through the inside. The milk flowed into ten-gallon aluminium churns with chamfered mushroom-shaped lids attached by chains. (The shape of the aluminium milk churn was derived from the old plunge butter churn.) Milk churns were phased out in 1979 – largely superseded by refrigerated bulk milk tanks and refrigerated tankers. By then there were fewer small farmers milking cows in out-of-the-way places, and fewer farms not connected to mains electricity.

I well remember how nearly every farm in our area when I was growing up had a milk stand at the roadside, from where the full churns were collected by the Milk Board and empties left for the next day. Full churns were heavy and had to be left on a raised platform at the same height as the bed of the milk lorry so that they could easily be transferred onto it. On hot summer days the milk would heat up in the aluminium and be none too fresh by the time it was collected. Every farmer was entitled to have his milk collected and paid for at the going rate, no matter how many cows he milked.

As the twentieth century drew on, farmers needed a quicker and less laborious milking system than individual units to milk the bigger herds that were coming in. Thousands of designs were patented all over the Western world for labour-saving milking parlours using whatever technology was available at the time. There were abreast parlours, tandem parlours and herringbone parlours, all variations on a design intended to allow as many cows as possible to be milked by one milker standing in a pit between two rows of cows with their udders at about chest level. This did away with bending down to attach the clusters, even though the time saved was at the cost of standing in a pit and being splattered with cow muck and urine.

In the 1930s, Henry Jeffers, an American dairyman, invented the rotolactor, a carousel with places for 50 cows – a bit like the London Eye laid on its side – which was timed to do a full revolution every 12½ minutes, long enough to wash a cow's udder and milk her, although the speed could be altered up or down. Rotary parlours involve less moving about for the operator because the cows come to him and step onto the moving platform themselves, enticed by the promise of cow cake. When she is finished milking, her udder is disinfected, the exit gate opens automatically, she walks out and steps off the platform and away.

Large rotary parlours will accommodate about 70 cows at a time and can milk up to 700 cows every hour. Everything is automatic. Each cow is fitted with a collar containing a programmed sensor, which gives the computer the necessary information: how much to feed her, the drying-off date when her lactation is finished, and when she last calved. It even automatically detects when she is on heat and has to be inseminated. But the really clever thing is that the information is fed into the system which then operates various gates for sorting the cows after they come out of the parlour. Those that need to be served, for example, will be directed automatically into one pen, those needing treatment will be directed into another, and so on. But the cows still have to be milked – up to three times daily – which can take up most of the working day.

So it was hardly surprising that some bright spark should invent the milking robot. This is an astonishing machine that uses every bit of modern electronic and computer equipment to milk about 70 cows a day without human labour. It is founded on the principle that a cow will decide when she wants to be milked and that being milked little and often will produce more milk than sticking to rigid milking times. Modern dairying has resolved itself into two competing philosophies, reminiscent of the differences that arose 20 years ago in baby-rearing: one maintaining that routine was the key to a contented infant, the other that baby-led feeding on demand would get better results. And just as with milking cows, each had its fierce devotees.

With robots, there is never a time when you don't need somebody on call; there might not be much to do, but you can never say that milking is finished. For this reason some farmers with large herds have tired of their robots and gone back to twice- (or thrice-) daily milking in large sophisticated parlours where the milker handles the cows every day, sees immediately if anything is wrong and deals with it. Cows in heat are spotted and separated, and lameness and other injuries are seen within a few hours. Having robots do the milking can easily lead to lazy cow management unless the herdsman spends more time with his cows than he does with his wife and children – as one disgruntled wife complained to me.

This is all a far cry from our house-cow. When we moved to Picket How we found ourselves a couple of miles from the nearest shop, through a couple of gates, across three fields, at the end of a rough track. So in a surge of youthful enthusiasm for self-sufficiency I asked my cattle dealer friend to find us a house-cow.

'Something quiet, easy to milk – and cheap to keep.'

A few days later a little cattle wagon came rocking over the potholes in the lane, turned into the yard and disgorged a slightly shaken-up fawn-like red-and-white heifer. She stood quietly in the yard looking around, sniffing the air and taking in her new surroundings.

My wife said, 'She's wondering what kind of place she's come to. I think we should call her Alice. Alice in Wonderland.'

And so that's what she became. She was a newly calved Ayrshire heifer, with nicely shaped and well-spaced teats just right for hand-milking. Or so I thought. She had been milked a few times, but never by hand and I was far from used to hand-milking. I found it tedious and annoying – and above all tiring – to have to spend twenty minutes at either end of the day extracting milk from this nervous little heifer. She didn't kick when I milked her, but she did not much like being milked by hand. She got fidgety towards the end if it took too long. And we found she gave far too much milk for our household needs. She was giving nearly three gallons a day, which is a hell of a lot of milk unless you're going to make butter or cheese, which neither my wife nor I could be bothered to do. The dogs got quite a lot of it, but there was still too much, so we got two pigs and turned them into bacon, which was very good stuff.

I persevered with her for a few months and she became quite well trained, coming to the gate at milking time and walking into the byre to be tied up, ready for her cake and standing to be milked. But she must have sensed my impatience with her and that I resented being tied to milking the cow no matter what else we were doing, every morning before breakfast and in the evening. She started a habit that eventually put me off hand-milking for good. As milking progressed she would flick her tail so that the hairy tip, with cow-muck hardened on the end, would lash me on the back of the head or round my neck or face. If there was soft cow muck or mud on her tail I would get splashed with the damned stuff. I started tying her tail on the other side of her body with a piece of baler twine, with a slip knot round her tail and a loop around her neck. I had to do this before I sat down on the milking stool because, if I didn't, as soon as I leant my head and shoulder into her flank to start milking she would flick that bloody tail.

She must have been irked that she no longer had the swishing of her tail as an outlet for her irritation, because she developed another trick that ended my relationship with her. She would allow me to take a pail of milk from her and just before I got to the strippings, she would lift up her hind hoof and plonk it into the full bucket. She didn't kick, she just waited for the right moment and expertly put her filthy foot into two or three gallons of hard-won milk, making sure it would be undrinkable. Yelling at her and slapping her just made her worse and she got to be so nervous that she became nearly impossible to tie up in her stall.

When she came into season I had my Hereford bull serve her and sold her in-calf, jolly glad to be shot of her. Poor Alice.

It didn't seem right to ask David Baynes, who produces Northumbria Pedigree Milk at Marleycote Walls near Hexham, if he had ever thought of giving up milking cows. But in fact he told me without my asking that he had been through many sleepless nights caused by the fear of debt, loss of his farm and that he would be unable to provide for his family. The Bayneses have farmed at Marleycote Walls since the 1860s. Their herd of Dairy Shorthorns have long pedigrees back to the 1930s and beyond, and are indigenous to Durham and North Yorkshire. More recently they have added Ayrshires to the herd because it was hard to find enough good Dairy Shorthorns. The average life for a commercial Holstein dairy cow is two to three lactations – that means they are about four or five years old when they are culled, worn out by making industrial quantities of milk – whereas some of the Bayneses' cows are into their tenth lactation and still going strong. This is a reflection of the ethos behind their farming. There is a tension in farming, as I suppose in any other creative activity, between doing it for love and doing it purely for money. These two are at either end of a scale, although one does not necessarily exclude the other. And somewhere in the middle, there is a point where happiness and success can be combined. That is what the Bayneses have achieved. But not without considerable effort, luck and risk.

Father and mother and two sons and their wives run the farm. One son lives for the cows and the other runs the processing, packaging and deliveries. It was over a decade ago that they realized they had to increase their income if they were going to be able to support three families from the farm. They faced a choice: go all industrial, enlarge their herd and replace the Shorthorns with Holsteins; or find a way of increasing the income from their milk. They did not want to change breed, largely because Shorthorns are well suited to the land and climate of the north-east of England, but also because their milk is so much superior to the 'whitewash' that Holsteins produce, and – they dare say it – they love their cows. So they pushed the boat out and built an airy, spacious shed for 160 cows, with wide passages and soft-bedded cubicles and automatic mucking-out, and installed two robotic milking machines. Everything was done on a generous scale, with much labour-saving incorporated into the design. I felt that if I were a cow, I would be pleased to live in this shed, my every need satisfied, warm in winter and carefree in the tranquil summer. It was light and bright and the cows ambled at ease below the viewing gallery, from which every part of the shed can be surveyed without disturbing its residents.

The milking robots cost 'the price of three decent cars each', said David Baynes, though he was reluctant to tell me exactly how much. (The current price is about £130,000.) They are manufactured by Lely, a Dutch company, and are astonishing machines that have transformed the milking of cows. There is no set milking time. The cows come to be milked by the machine when they feel like it, or they come for their ration of cattle cake and get milked. Their ration is measured by the computer, which has read the transponder in their collar, and is dispensed into a trough in front of them as they approach. If a cow has recently been milked and is only coming for cake, the computer will register this and she will get nothing. Some dominant or greedy cows refuse to move out of the way to let others in even though there is nothing for them to eat. In most cases they eventually get the message and move away, but if there is a hold-up, the computer will send an alert to whoever is on duty to come and sort it out.

As the cow is standing eating her cake, an arm swings down beside her and holds her in position. Her udder is washed automatically by a spray, and then another arm swings under the udder, locates the teats and attaches a rubber cup to each one. If the cow kicks the cluster off – which occasionally happens – the machine will repeat the action until the cups are firmly attached and the cow begins to let down her milk. Each quarter is milked separately, so that when a quarter is dry, the machine will stop the vacuum to prevent harm.

The robot analyses the quality of the milk in each quarter and if for any reason it is not good enough to be conveyed into the bulk tank for bottling (because it is contaminated), it is automatically diverted away to be disposed of. Colostrum for calf-feeding is directed to a separate storage tank. The system can detect illness in a cow, by reading her temperature or weighing her daily to see if she is losing weight. Mastitis, for example, can be spotted 48 hours before there are any obvious symptoms.

When cows are at grass in the summer, the system can be programmed to direct them to a particular paddock, and if any cow has not been milked, she will be sent back into a holding pen to go round again.

The Bayneses' milk is pasteurized and bottled, either as whole milk, or skimmed with the cream bottled separately, under the Northumbrian Pedigree label. They deliver it themselves directly to smaller retail shops within a 60-mile radius of the farm and it sells at 90p a litre, in contrast to the farm-gate wholesale price, which currently hovers about 30p.

Keen to capitalize on its quality, they maintain the naturalness of the milk by not homogenizing it, although they do blend the evening's milk, which is higher in butterfat, with the morning's. The emphasis is on feeding the cows as much of a home-grown (GM-free) diet as possible, based on grass silage with whole-crop wheat (cut and made into silage while it is still green), barley and some soya and rape meal.

The Bayneses' enterprise is a model of how a farmer can become independent of the control of dairy wholesalers, processors and, above all, supermarkets, and get a proper price for their milk. It is not easy to find and sustain a retail market and there is no doubt that without small shops they would have been unable to prevail against the power of the supermarkets. It is crucial that they deliver punctually and reliably and they put great effort into it. Their refrigerated vans cover huge distances every week, and hardly ever fail to get through, even in the worst of weather. The responsibility is relentless, from perpetual care for their cows, to satisfying their customers. They have few days off, but like most farmers, I suspect they can't see the point when, like an artist, their life and work are so entwined. Where would they go when everything that gives meaning to their lives happens at home? The Bayneses are one of Edmund Burke's 'little platoons' upon whom we depend to put food on our tables. As Adam Smith rightly observed: 'It is not from the benevolence of the butcher, the brewer, or the baker, that we expect our dinner, but from their regard to their own interest. We address ourselves, not to their humanity, but to their self-love, and never talk to them of our own necessities but of their advantages.'

CHAPTER 5

The Channel Island Breeds

W ELL INTO MODERN times, there were three distinct
Channel Island breeds, Jersey, Guernsey and
Alderney, which tended to be lumped together and
called Alderneys. The last of the true old Alderneys, the result
of centuries of selective breeding for rich creamy milk, were
killed and eaten by the German forces occupying the island in
1944. When A. A. Milne refers to the Alderney in 'The King's
Breakfast', he is almost certainly using it in the old sense of a
composite for the Channel Island breeds; the cow was most
likely a Guernsey:

> *The Dairymaid*
> *She curtsied,*
> *And went and told*
> *The Alderney:*
> *'Don't forget the butter for*
> *The Royal slice of bread.'*
> *The Alderney*
> *Said sleepily:*
> *'You'd better tell*
> *His Majesty*

That many people nowadays
Like marmalade
Instead.'

When the fashion first arose for its importation, the Alderney was derided as being a cow for a gentleman's park: 'it is thought fashionable that the view from the breakfast or drawing room ... should present an Alderney cow or two at a little distance' to provide rich milk for his lady's tea table.

The modern representatives of the Alderney are the Guernsey and the commoner Jersey, both almost certainly collaterally descended from the ancient yellow and broken-coloured stock that populated the coast of north-west Europe from the Low Countries to Brittany. The French breeds, the Isigny from Normandy and the Froment du Léon from Brittany, are their close relatives. For many centuries the cattle of the Channel Islands had been isolated from both France and England and their distinctive breeding jealously protected by the islanders from adulteration by imported cattle. As early as 1789, it was made unlawful to import any cattle into Jersey, partly to protect their value, because French cattle were being imported into England via Jersey and reducing the value of the native beasts. A similar prohibition was imposed in respect of Guernsey in 1819. The Jersey embargo was only rescinded in 2008. There has long been a healthy trade in exports the other way, from Jersey to England, contributing substantially to the island's income. Channel Island cattle were found in southern England as early as 1700. The historian of the Jersey breed, Eric J. Boston, writes that 'the sloop, *Jane*, of Guernsey, was chartered on the 1st of September 1741, to proceed to Jersey and take on board eight cows for Southampton'.

There are records of the breed's importation since at least 1724, largely by gentry and noblemen who fancied them for their rich milk and to adorn the parkland around their seats. During the eighteenth century, the annual importation was about 1,000 cows and bulls. This increased in the nineteenth century to between 1,500 and 3,000 a year. In July 1819, it was reported in *The Times* that shipments to England 'had completely drained the Islands' of cattle. In 1834, the Jersey cow was distinguished from the other two by the Royal Jersey Agricultural and Horticultural Society, driven by Sir John Le Couteur, a decorated soldier during the Napoleonic Wars who retired from the British army to his native Jersey and took a great interest in the breed. The society drew up a scale of breed points to encourage high standards. During the nineteenth century, Jerseys spread into every part of the kingdom: from Devon, where they were crossed with the native Devon breed to produce the South Devon; to Ayrshire, where both the 'Alderney' and the Guernsey contributed to the development of the Ayrshire breed; and into Ireland, where much of the Alderney blood found its way into the 'poor people's breed of little mountain or Kerry cow' in County Cork.

It is possible that the breed arose many centuries ago from a cross between some Norman import and a little native type in the Channel Islands that resembled the older forms of Kerry, Cornish, Welsh, Shetland and Ayrshire cattle that accompanied their Brythonic owners retreating before Germanic and Norse invasions from the east. One commentator makes the point that these breeds were all found in places where there are prehistoric remains, and suggests a close connection with *Bos longifrons*, the little 'Celtic cattle'. Trow-Smith dismisses this theory, saying that there is evidence of Alderney blood being introduced into the native cattle of these Celtic places. But in the light of recent research into the migrations of Brythonic people, even before, and certainly after, the Roman occupation of Britain, it would hardly be surprising if they took their cattle with them.

There was a widely held belief that Alderneys were not hardy enough to survive in the English climate under ordinary conditions of husbandry. George Culley thought the 'breed too delicate and tender ever to be much attended to by our British farmers', but he could not deny the richness of their milk and the fineness of their flesh. Their extreme dairy characteristics make them look fragile, and there is some truth that the calves are susceptible to cold and wet; they have a greater surface area of skin relative to their body mass than other larger cattle, but they are hardier than they look.

The Jersey's value is twofold: their milk is high in solids, up to 4.9 per cent butterfat and 3.8 per cent protein; and they can be grazed more intensively than larger cows. In measuring the relative merits of breeds, it is the weight of produce that they give from an acre of land rather than the yield of individual cows that counts. Using this measure, the Jersey comes a long way up the scale. Brown Bessie, from Orfordville, Wisconsin, was a champion Jersey 'butter cow' that averaged over 40 lb of milk a day, making 3 lb of butter every day for the five months of the Chicago World's Fair in 1893, held to celebrate the four hundredth anniversary of Columbus's landing in the New World. Mainstream Barkly Jubilee holds the record (twice) for Jersey milk production. She is the first Jersey to give over 50,000 lb of milk in a lactation. She gave 49,250 lb in her second lactation in 2006, after calving at three years and six months, and 55,590 lb in her third, after calving at four years and eight months.

It is not only modern Jerseys that produce a lot of milk. Lily Flagg, reared in Northeast Huntsville, Alabama, was champion butterfat and milk producer of 1892 when she produced 1,047¾ lb of butter and 11,339 lb of milk. Her owner threw a famous party in her honour, painted his house butter yellow in tribute and described her as 'a cow worth kissing'.

Jersey (and Guernsey) milk is what used to be called 'gold top': high in butterfat, protein, minerals and trace elements and a much richer, creamier colour than other milk, due to the high concentration of beta-carotene, which the cows extract from the grass. It has 30 per cent more vitamin D than other milk. The best cows produce up to 30 litres of milk a day, with the average Jersey in the UK producing about 6,000 litres in a lactation.

The effect of preventing the importation of cattle onto the island, apart from maintaining the breed's purity, was to protect dairy producers and ensure continuity of supply to the islanders. In 1954, the States of Jersey set up the Jersey Milk Marketing Board, a producer-owned cooperative that buys and processes all the milk produced by all but one of the island's 24 farmers. In 1981, the Board established a commercial arm to sell its dairy products under the brand Jersey Dairy. It is run, as the old MMB was, as a monopoly, finding a balance between paying the farmer a proper return for his effort and costs, and charging a fair price to the customer.

It can do this because Jersey is functionally independent from both the EU and Britain and is free, to an extent, to make its own rules. The Jersey Milk Marketing Board has the characteristics of the ancient idea, decried by modern free-market economists, of the 'just price'. It is interesting to contrast the effect on Jersey farmers and consumers of using the Board to establish a fair price, with the free-market experience of dairy farmers across the Channel in the UK. The principle of the just price was attractive to medieval Christian scholars, prominently Thomas Aquinas, who built on the ideas of the ancient philosophers, trying to strike an economic balance between a producer's monopoly, which has the consumer at his mercy, and an unregulated market, which leads alternately to gluts and shortages, and imperils a reliable supply.

The result for Jersey is that its dairy farmers are not going out of business in droves; rather they are making a secure living that gives them confidence in the future. They do not have to increase the size of their herds, or try to put their neighbours out of business to further their own, nor do they have to become involved in retailing or processing their milk. Instead they can concentrate on looking after their cattle, which is what they do best.

On the customers' side, there is a reliable source, at a reasonable price (currently about £1.10 a litre), of fresh clean milk that has not travelled further than a few miles across the island. The land is not being ravaged by large amounts of artificial fertilizer and chemicals, and there is not one enormous industrial mega-dairy with cows that are never allowed outside. From the cows' point of view, they live longer and they have the sun on their backs.

The Richardson family have been milking Jerseys since 1925 at Wheelbirks Farm, just south of Stocksfield in Northumberland. They make Jersey ice cream as a commercial response to the poor price of liquid milk and have an increasing market for unpasteurized Jersey milk. Wheelbirks is probably a corruption of the Old English *weald*, meaning 'open country', and *birks*, which is Old Norse for 'birch'.

The Richardsons were Quakers, originally from Whitby, who in the eighteenth and nineteenth centuries had a considerable tanning works on Tyneside. David Richardson, the great-grandfather of the brothers who currently own the farm, bought it in the 1880s and set about turning it into a gentleman's estate that he could use as a retreat, for his own family, but also for his staff and their children. He rebuilt the farmhouse and cottages, set out watering places in the fields, built capacious modern farm buildings and bridges, and about the turn of the century even constructed a sanatorium for his tannery workers, who were prone to contract TB. It's built on stilts over a gully in a cruciform shape, with walkways like drawbridges leading to it from the higher ground around. It was never used as a sanatorium, partly because it was never finished. Finney Seeds of Tyneside used it for a while as glasshouses for testing their seeds and growing cut flowers for the local city market.

During the course of the renovations and building work on the estate, Richardson had aphorisms and verses carved onto stones built into the walls. Over the cow byre, now turned into a milking parlour, is inscribed the enigmatic line: *Be sure your work is better than what you work to get.* Dere Street, the Roman road from York to Corbridge, crosses the farm, and where the ancient road dipped down to Stocksfield Burn, Richardson built a fine stone bridge, on the parapet of which appears the first line of a poem by Christina Rossetti, 'Up-Hill':

Does the road wind up-hill all the way?
  Yes, to the very end.
Will the day's journey take the whole long day?
  From morn to night, my friend.

In the wall of one of the cottages is *A stone that is fit for a Wall will not be left in the Way.* On another wall, enclosing some wasteland, is the exhortation to mankind in Genesis *Be fruitful and multiply; Replenish the earth and subdue it.*

Two large branches of dried-up holly hang from the rafters in a cattle shed at Wheelbirks. Hugh Richardson told me it was male holly (it has to be male, apparently) to counteract ringworm. Apparently the spores of the fungus prefer the holly and leave the cattle alone. Ringworm is a highly contagious spore-forming fungus *Trichophyton verrucosum*. It affects all species of mammals, including cattle and man, and forms circular scaly lesions on the skin. The scab in the centre of the lesion tends to fall away, leaving a series of rings. It was once common when cattle were housed in old buildings, where the spores can survive for years in woodwork and crevices. The animals would become infected when they were brought in for winter, and unless they were treated, the ringworm would stay with them until the following spring; when they were turned out to grass, it was killed quickly by sunlight. Young cattle are worst affected because some degree of immunity is conferred by becoming infected.

I once asked a French farmer whose calf sheds were festooned with holly branches if it worked. He replied, 'I don't know, but we've had no ringworm on this farm since we started hanging up holly over ten years ago.'

The Richardsons use genomic sex-selective insemination for the first AI service on all their cows to ensure that all the calves born are female, because Jersey bull calves are almost worthless unless they're wanted for breeding. If a second insemination is needed, they use a Belgian Blue to try to put some beef into them. The calves are then sold on at three months or so for beef or veal. The herd is closed, which means that no stock is brought onto the farm. The clever way of breeding bulls is to use fertilized male embryos from selected cows and bulls with desirable characteristics and implant them into their own cows. The resulting bull calf will not be related to any cow on the farm and can be used to run with the herd to sweep up any that do not take with AI.

Each cow wears a device on her ankle, like the tags the courts sometimes make criminals wear to keep track of their movements; it records the number of steps she has taken during the day, on the principle that a cow in season will be restless, moving around when the rest of the herd is lying down or cudding or just quietly grazing. A signal from the anklet will trigger an alert when she comes in for milking, at which point she can be separated from the herd for artificial insemination.

The sexes are more defined in dairy than in beef cattle. The bulls are more masculine and aggressive than beef types and the cows more feminine. Jersey bulls are notoriously bad-tempered and have to be treated with a good deal of caution. What they lack in size they more than make up for in attitude. I could hear the bellowing of the young bull running with the dry cows at Wheelbirks before we reached the field with a big bale of silage on the tractor loader. When Hugh Richardson set down the half-ton bale to remove the plastic netting, the bull set about trying to demolish it, scattering silage all around and rolling it across the field, bellowing all the while. Hugh only managed to stop it by lifting what was left of the bale into the feeder with the loader.

The delicately feminine Jersey cows, on the other hand, are docile, sometimes nervous, and, even for cattle, inquisitive. They ambled up the field and crowded round; some tried to lick me, others, a bit pushier, rubbed their heads against me, and some stood and stared, trying to work out what this stranger was doing in their field. The best of the cows have the classic wedge shape, being deeper in the rear half, which is the business end. As Jerseys only weigh about 400 kg, they can be left to graze the fields longer in the autumn because they won't cut up the fields as much as heavier breeds. They also tolerate the summer heat well.

The femininity of the cows and masculinity of the bulls seems to emphasize a kind of dualism in genetics that balances a particularly prominent characteristic in one animal by making its converse equally prominent and so creating the whole through a unity of opposites.

Jerseys tend towards a general shade of fawn, although they come in all shades of brown, from light tan to mahogany, almost black. Pure-bred animals have a lighter circle around the muzzle and eyes, a dark switch – the hair at the end of the tail – and black hooves. In the mid nineteenth century there was a fashion for silver-grey Jerseys, and a later vogue for whole colours as opposed to 'broken' or mixed colouring. Rich owners started to take an interest in showing their stock, and in 1879, at the London Dairy Show, Jerseys comprised the single most numerous breed, with 253 entered in the various classes. No other breed aroused such interest as an object of fashion.

Jerseys have a greater propensity than most breeds to suffer from milk fever, called hypocalcaemia, caused by a lack of available calcium in the blood, typically around calving. Unless the cow is quickly given intravenous calcium, magnesium and glucose, she will go into a coma and die. Fortunately the remedy promotes a miraculous recovery and she will be back on her feet within half an hour as if nothing had happened. Before intravenous injections were possible, the remedy for milk fever was to inflate the cow's udder through the teat with a bicycle pump. I suppose they were trying to push back some of the calcium into the bloodstream, or something like that. It's hard to imagine how anybody thought of doing such a thing.

There has been a remarkable reversal of fortune between the Jersey and its cousin, the other Channel Island breed, the Guernsey. In 1955, Guernseys comprised 5.3 per cent of the dairy cattle in England and Wales – about 130,000 cows. There were fewer Jerseys. In the last 60 years, the Guernsey population has dwindled to about 5,000 cows in the UK, while the Jersey has more than held its own.

Guernseys are bigger, stronger-looking cows than Jerseys – about 50 kg heavier – less deer-like and doe-eyed, and they have never been quite so subject to fashion as their cousins. They are mostly a yellowy fawn, a proportion broken-coloured, with a white belly, tail and margin to the muzzle and distinctive amber hooves. Their colour shows a closer relationship to the European Blonde cattle races than the Jersey and may also reflect a discrete crossing of improved Shorthorn in the early nineteenth century that increased their size and improved their carcase shape, making them more dual purpose than the delicate Jersey. Both Channel

Island breeds descend, a long way back, from a sub-type of the Blonde races, of which the ancient Brown Swiss and the French beef (originally draught) breed, the Blonde d'Aquitaine, are the most numerous modern representatives. All Channel Island cattle were kept for draught as well as dairy and meat production. But it is their unique genetic inheritance from African and Asian cattle and preserved by long isolation, that sets the Channel Island cattle apart from every other British breed.

The Channel Islands were part of the kingdom of Normandy and under the suzerainty of the dukes of Normandy. In AD 960, Duke Robert settled some monks on Guernsey who brought cattle with them from mainland France, most likely from Brittany. A hundred years later, a further monastic settlement imported brindled cattle from the mainland and the Guernsey was on its way to the distinctive breed it is today.

The Jersey breed was kept isolated and pure for 250 years, from 1763 to 2008, because the States of Jersey closed the island to prevent any genetic adulteration. Guernsey did the same from 1819. However, it is recorded that zebu cattle were crossed with Channel Island and Devon cattle on the mainland between 1795 and 1805.

The Guernsey's main attribute is that it produces milk unlike that of any other European breed, ideally suited to cheese production and giving health benefits that are only now becoming better understood. It has more solids than other milk, which makes for a firmer curd, and it contains 15 per cent more calcium, high levels of beta-carotene (vitamin A) and 33 per cent more vitamin D.

Despite the efforts of the processors and supermarkets to standardize it, not all cow's milk is the same. Its taste and quality depend on the season of the year, what the cows have been eating, the stage of their lactation and even the weather. But there is another difference. Although most milk is water, the valuable part is in the roughly 15 per cent solids: butterfat, which varies with the breed and feeding and is between 3.2 and 5 per cent; protein, which is between 3 and 4 per cent; milk sugar (lactose) and minerals. The protein fraction is roughly 80 per cent casein and 20 per cent whey and is the part from which cheese is made. An increase of 0.1 per cent in the protein content of milk gives a 3 per cent gain in the amount of cheese produced.

While casein coagulates with the addition of the enzyme rennin, whey is unaffected and remains liquid unless it is altered by heat and acidity. Rennin is synthesized in the stomachs of mammals in the first few weeks of life, where it causes the maternal milk the young animal consumes to curdle so that it can digest its proteins. This is essentially the same process that has been replicated in cheese-making since the beginning of time. Originally the rennin came from cells in the dried stomach of a young calf, but as there are not enough young calves to go round these days, it is now mostly made synthetically.

Of the five types of casein in milk, kappa-casein is the key protein in cheese-making. It has three main alleles (genes), Kappa A, Kappa B and Kappa E, and each cow inherits one of these from each parent. So a cow with a double inheritance of Kappa B (code BB) produces milk that clots 25 per cent faster and will make 10 per cent more cheese that is twice as firm as that from a cow with an AA gene. The kappa-casein BB allele occurs most often in traditional breeds of cow, and the Guernsey stands out among them because 60 per cent of the cows carry the Kappa B gene.

But Guernsey cows are also unusual in that 96 per cent of them give milk containing another protein, beta-casein A2. When the beta-casein chain of protein was first analysed, it was given the code A1. Only later was it discovered that not all beta-caseins are the same. Some were found to have proline at the 67th amino acid in the chain, whereas in A1 milk it is histidine. Researchers also discovered that the A2 chain containing proline was the original and the A1 a genetic mutation that they suggested had occurred a few thousand years ago in European cattle.

This mutation is commoner in the big black-and-white breeds of northern European descent, such as the Holstein and Friesian, which are the breeds that produce most of the milk in northern Europe (excluding France), America, Australia and New Zealand. Friesians and Holsteins usually have alleles for A1 and A2 in equal proportion, whereas Jerseys and other traditional European breeds have about a third A1 and two thirds A2. Guernseys are unique in having only 10 per cent A1 and 90 per cent A2.

Research by Professor Keith Woodford in New Zealand into the structure of beta-casein (explained in his book *The Devil in the Milk*, published in 2007) found that A1 and A2 beta-caseins react completely differently with enzymes found in the digestive system. When people try to digest A1 beta-casein, a small protein, beta-casomorphin-7 (BCM-7), can be released. It is BCM-7 that Woodford describes as 'the devil in the milk' because epidemiological research and animal studies in the 1990s in New Zealand found a correlation between consumption of milk with A1 beta-casein proteins and certain chronic diseases, such as Type 1 diabetes, heart disease, even schizophrenia and autism. The reason, he suggests, is that BCM-7 is strongly bonded to proline in the A2 milk, which prevents it from being released into the system, whereas histidine has a weak bond with BCM-7 and is easily released and imperfectly digested as it passes through the human intestine. BCM-7 is an opioid that doesn't occur naturally in the human body, and when released, he suggests, it interferes with the human digestive system and affects the internal organs and brain stem.

These findings prompted a group of entrepreneurs to set up A2 Corporation, to promote A2 milk as the healthy alternative to regular A1 milk and even (unsuccessfully) petition the New Zealand Food Standards Agency to require A1 milk to carry a health warning.

In 2009, the European Food Safety Authority (EFSA) reviewed the research (but did not undertake any of its own) and refused to find any connection between chronic disease and consuming A1 milk. This has not deterred a significant number of people from acting as if Woodford's findings are true. They point out that most mammals, including goats, sheep and humans, produce A2 milk, as do traditional African and Asian cows, water buffalo and yak. Only those breeds that have been 'contaminated' by interbreeding with cattle carrying the genetic mutation produce 'unhealthy' A1 milk.

The older European breeds like the Brown Swiss, Guernsey, Jersey, Montbéliarde, and Gloucester have quickly come to be known as A2 breeds (although their milk is not exclusively A2), while the higher-yielding types with the genetic mutation, particularly the black-and-white breeds, are loosely referred to as A1 breeds, although they will still produce *some* A2 milk. These include the Holstein, Friesian and Ayrshire.

A group of Russian researchers took it further and found that BCM-7 passes undigested through the gut wall into the blood of babies fed infant formula, causing delayed brain-to-muscle development. Another report, published in the Indian *Journal of Endocrinology and Metabolism* in 2012, agreed with Woodford's findings that A1 milk is a risk factor for certain diseases and psychological disorders. His opponents, which include most of the large dairy processors and modern industrial farmers, say the findings are little more than speculation because most of the research was done on animals, the diseases concerned have many contributing causes, and there is little evidence that animals suffer from autism or schizophrenia, which probably do not have an organic cause anyway.

One researcher, however, Dr Natasha Campbell-McBride, has come up with Gut and Psychology Syndrome (GAP or GAPS), a term she has trademarked, which describes a connection between the functions of the digestive system and various illnesses and behavioural disorders, such as autism and psychiatric afflictions. If you do drink milk she recommends it be unpasteurized and organic. And presumably the best option is to get it from Guernsey cows.

CHAPTER 6

The Black-and-White Revolution

THERE IS A story, partly verified by Tacitus and Pliny the Elder, that sometime about 100 BC, the Chatti tribe, occupying lands in Hesse, fell out amongst themselves. Unable to reconcile their differences, a portion of the tribe left to live elsewhere, travelling with their black cattle further west, to the shore of the North Sea, where they settled on the island of Batavia, in the delta of the rivers Rhine, Maas and Waal. Their neighbours were the pastoral Frisii tribe, whose cattle were pure white. In time the two strains interbred, creating a black-and-white type that came to be called the Friesian, renowned for its excellence. So valuable were the cattle that rather than the Batavians being conscripted into the service of the Empire, Rome considered that they would be more useful if they concentrated on cattle husbandry and paid tribute in ox hides and horns. For over 2,000 years they maintained the purity of their breed, which became renowned all over northern Europe for its remarkable capacity to produce large amounts of butter and cheese from the fertile alluvial soils of the polders.

This black-and-white 'Dutch' type has been known in England since at least the sixteenth century. From time to time they were imported into the east of England, as fashion and need dictated. A wave of importations during the 1880s meant that by 1900 there were 30 or 40 herds scattered around England. They were known as Holsteins, Frieslands or Friesians and were the kind of dual-purpose cattle – beef and milk – that filled the gap in the market that the original Shorthorn had once occupied. The breed was promoted by some able, rich and enthusiastic breeders, notably the Strutt family at Terling in Essex, who had started dairying on land abandoned by their tenants during the agricultural depression after 1875.

Early breeders would not accept any red-and-white animals, but as these were often exceptional milkers, breeders who were reluctant to abandon good cattle just because they were not black and white formed their own society in 1951. This eventually merged with the black-and-white British Friesian Cattle Society in 1985, and in 1999, the Holstein and Friesian societies merged to become Holstein UK. In reaction, in 1990 the British Friesian Breeders Club was formed to try to preserve the original dual-purpose, grazing character of the British Friesian, which was being lost in the Holstein-based rush for ever-greater yields of milk.

Over the decades, American and Canadian breeders, chasing milk yield, had moved away from the dual-purpose ideal and concentrated on ever more specialized dairy animals with carcases markedly less able to make beef. To deal with the fact that the male calves are worth very little, they either use sex-selected artificial insemination to get female calves, or simply kill the bulls shortly after birth. Thirty years ago, when I used to buy week-old calves for rearing, I felt sorry for the little waif-like dairy bull calves brought into the auction and sold for next to nothing. Some only made £1 because they were not worth the cost of rearing. Many went straight for slaughter, their carcases to be ground up for pet food or fertilizer or animal feed. That is the dark side of breeding extreme dairy types for milk yield.

American farmers took to the Holstein, glorying in its capacity to produce more milk than any other dairy breed in the world. President William Howard Taft even kept a Holstein cow, Pauline Wayne, as the official presidential pet between 1910 and 1913, grazing the White House lawn and providing milk for the First Family. The breed's tremendous production has been achieved without concern for the longevity of the cows. The average Holstein cow in the US now lives for less than five years, with an average of fewer than three lactations. Farmers compete to breed a cow that gives the greatest yield in a single lactation. The current US Holstein record is held by the rather inelegantly named Bur-Wall Buckeye Gigi EX-94 3E, which produced 74,650 lb (about 9,000 gallons) of milk in 365 days in 2016. (A gallon of milk weighs between 8.5 and 8.8 lb depending on the density.)

Across the world, from Canada to Russia, India to New Zealand, the Holstein is, by a significant margin, the highest-yielding commercial dairy cow, with a breed average of 18,500 lb (2,200 gallons) in a lactation. The fat content (about 3.7 per cent) and protein (about 3.2 per cent) is lower than traditional dairy breeds, but that is hardly surprising given the sheer volume of liquid the cow excretes over its short life. Individual Holstein cows are capable of phenomenal yields. Milking ability is roughly half inherited and half down to the feeding and general care of the cow. The half attributable to feeding is only achieved by heavy feeding of

English Longhorn cows at Calke Abbey. Until the eighteenth-century agriculturalist Robert Bakewell ruined their milking capacity in the search for a beef carcase, Longhorns were the quintessential triple-purpose English cattle – traction, milk and meat – and found in nearly every part of the kingdom. Despite their formidable appearance, they were docile and gave a respectable yield of rich milk, ideal for cheese-making. They could turn almost any vegetation to their advantage over a long life, at the end of which they readily finished for the butcher.

Gloucester cow and calf showing the characteristic finching along the belly, up the tail and halfway along the back. It resembles the Kerry with its lyre-shaped horns and delicate dairy frame.

Gillray's 1802 cartoon, 'Vaccine-Pock hot from ye cow', lampooning the widely held fear that introducing bovine tissue into human bodies would cause them to grow body-parts of cattle. Note the painting of the Golden Calf on the wall.

The Baynes's pedigree Shorthorn and Ayrshire dairy cows in their spacious, comfortable, airy quarters at Marleycote Walls in Hexhamshire. Note the slats in the concrete floor through which the muck and urine falls, thereby keeping the wide passageways clean. The muck is stored in underground tanks for spreading as fertilizer on the cropping fields. Here the cows are inside for the night having been out to graze during the day; they move around as they please, making their way to one of the milking robots when they feel the need to be milked, or fancy a feed of cake.

The eighteenth-century sculpture built into the outside wall of the east transept of Durham Cathedral. It records the legend of the founding of Dunholme – Durham as it now is – and shows the Dun Cow, her milkmaid and the woman who directed St Cuthbert's entourage to the place where the saint's body should lie.

The classic wedge shape of a superior dairy cow, with a capacious, well-shaped udder and medium-sized, well-spaced teats. Note the bulging milk-vein under her deep belly. She is from the Richardson's Jersey herd at Wheelbirks in Hexhamshire.

Slender feminine Holstein maiden dairy heifers before they have had their first calf.

A Luing bull calf on his home territory on the Isle of Luing. Emanating from a cross between a Highland cow and a Beef Shorthorn bull, this breed is marvellously adapted to the climate and terrain of the north and west of Scotland, where its thick coat keeps out the cold and sheds the rain. It can extract energy from the poorest herbage to grow a superior beef carcase.

A classic type of Hereford bull, just like my errant bull Jason. Note the finching running under the belly, up the dewlap and neck, to the characteristic 'bald' white face, the hallmark which the Hereford stamps on every breed it is crossed with.

South Devon bull and cow. The Guernsey inheritance is evident in their creamy skin, while the beefy shape and docile temperament comes from their Devon ancestors.

A breed 'beautiful in the highest degree' and unspoiled by the eighteenth-century improvers because it was unimprovable. This young Ruby Red Devon bull is from William and Richard Dart's herd at Great Champson, which is founded on some of the oldest and best Devon bloodlines.

To Arthur Young Esq: FRS &c&c&c for the Annals of Agriculture
The LONDON Cutting Names & Proportionate Price of Pieces
A West Country or any of the Fine flesh'd Sorts.

GOOD OX.

	Hind Quarter	D	F	No.	Fore Quarter	D	F
1	Sir Loin	3	5	9	Fore Rib containing five	5	6
2	Rump	5	6	10	Middle Rib containing four	4	6
3	Edge Bone	4		11	Chuck containing three	3	6
4	Buttock	4	6	12	Shoulder or Leg of Mutton Piece growing on the Chuck and part of the Brisket	3	6
4	Mouse Buttock	3		13	Brisket	4	.
5	Veiny Piece	4		14	Clod	3	.
6	Thick Flank Part growing under the Fat of the Buttock	4	.	15	Sticking Piece Neck End growing under part of the Clod	2	.
7	Thin Flank	4		16	Shin	1	2
8	Leg	1	6				

NB. The above supposed about Christmas as the fairest Season for Valuation.

This diagram from 1800 shows London butchers' cutting names and proportionate prices of a West Country (Devon) ox, 'supposed about Christmas as the fairest season for valuation'. It is addressed to Arthur Young 'FRS etc. etc. etc' for inclusion in the *Annals of Agriculture*.

2. The cutting names and prices of joints of a West Country (Devon) ox of about 1800. The diagram showed how the higher-priced cuts lay in the rear half of the beast.

A 'vast plateau of roast beef' with 'beef to the root of the lug'. This Aberdeen Angus bull from Andrew Elliot's herd at Blackhaugh, Clovenfords, Galashiels, shows the best of modern Scottish beef breeding. Note the small head relative to the deep, long, square body, with weight in the hind quarters where all the valuable cuts of meat are to be found.

Prize-winning young Cumberland White Shorthorn bull being made ready for sale at Carlisle market. The first cross with a Galloway cow produces the wonderful Blue-Grey, a superb hybrid suckler cow for marginal land. The breed is local to the hard moorland of the Scottish Borders and a testament to the instinctive skill of stock-breeders in the Border country.

Galloway cow with Blue-Grey calves, the offspring of a Cumberland White Shorthorn bull that appears in the photograph above. The heifers fetch a premium for their hardiness, longevity and capacity to rear a fine beef calf from some of the poorest land in Britain.

high-protein foodstuffs no cow would ever encounter in a natural grazing life.

Most of the protein now fed to cattle in the UK is from soya beans grown and shipped in from abroad. This has contributed to a surging worldwide demand for the 'king of beans'. Eighty per cent of the world's crop is grown in the US, Brazil and Argentina, and the acreage has increased 15-fold since the 1950s, with Brazil increasing production from 1.5 to 50 million tonnes on 30 million acres of what was once temperate savannah and rainforest. Whole communities have been displaced by the few huge corporations that control this highly mechanized monoculture. Hundreds of thousands of small farmers and their families have been induced to give up their land, with many reduced to the status of day labourers, while others have moved to the cities or are squatting in parts of the forest they have cleared for subsistence farming.

The crop in Brazil can only be grown by irrigating it with vast quantities of water drawn from the depleting reserves of the Guarani Aquifer and spraying it with large amounts of oil-based chemical fertilizers and herbicides. These are leaching into and contaminating the ancient underground body of fresh water. Perhaps more worrying is the routine application of the wonder herbicide glyphosate to soya plants genetically modified (GM) to be immune to its effects. Glyphosate kills all plant growth other than the GM soya. Some residue remains in the beans and the resultant oil, but it is claimed by its manufacturers to be harmless. The EU doesn't agree and is considering restricting its use or even banning it because it is feared it is carcinogenic. In North and South America, by contrast, there are no restrictions on its use; they rely on it too much to give it a bad press.

Ninety-four per cent of soya beans grown in the US in 2014 were from GM seed.

An example of the scale of the global trade is the huge importation at the end of November 2016 into Teesport, where Glencore landed 54,000 tonnes of 'soya feed products' shipped from their soya bean 'crushing facility' in Argentina, which is capable of processing 21,000 tonnes of beans every day.

Soya is rapidly becoming an essential ingredient in a considerable range of human foodstuffs and industrial products, including pesticides and textiles, although its primary use is in industrial farming of poultry and pigs, with a lesser amount used in feeding intensively kept dairy cattle, whose productivity depends on the protein obtained from it.

Glyphosate made a fortune for Monsanto, the American multinational company that first synthesized it. It is sold under various brand names – in Britain as Roundup. In 2008, scientists working for the US Department of Agriculture (USDA) described it as a 'virtually ideal' herbicide, 'a one in a hundred-year discovery, as important for global food production as penicillin is for battling disease'.

This assumes that the only way to feed people is through intensive industrial farming, with fewer and fewer farmers managing ever-larger acreages of monoculture sustained by agrochemicals. At the very least this assumption is questionable. Of course scientists involved in the manufacture of glyphosate are going to promote their product. But it is a false analogy to equate antibiotics used specifically to treat an infection to save life with a product like glyphosate that is designed to kill all life except the kind that scientists have created. It is far from certain that the routine use of such a substance will not be harmful in ways that these clever scientists cannot foresee. Already plants are developing resistance to glyphosate, just as bacteria have with antibiotics. Instead of putting our trust in and working with nature, increasing dependence on ever more ingenious ways of defeating natural processes may have rather different consequences from those the scientists hope for.

Monsanto claim that glyphosate is harmless once it reaches the soil, with negligible residue in plants. But foodstuffs are not tested for glyphosate by the US Food and Drug Administration (FDA). This is despite field tests finding it in lettuce, carrots and barley up to one year after the soil had been treated with it. The US government has set the acceptable daily intake of glyphosate at 1.75 mg per kilo of bodyweight per day, while the EU considers 0.3 mg per day to be the maximum safe intake. The truth is, nobody knows what is safe; it depends who is setting the limits. Not only are huge profits generated in the USA from worldwide sales of glyphosate, and above all the GM seed they have developed, but the herbicide is used so extensively and routinely that to set any lower limit on safe consumption would be to condemn vast areas of crop-growing land across the US as unsafe for cultivation.

Another concern is that in processing soya beans to make meal and oil, the bean is 'cracked' and the oil extracted chemically using hexane, a petrochemical solvent, which is added to the crushed beans. Hexane is a known neurotoxin and air pollutant. Tests have found residues of it in soya products, even though the processors claim almost none of it finds its way into the resultant oil or meal, and anyway it is harmless. Again, it depends who is doing the testing.

There is also some unease in certain circles over evidence that soya and anything made from it affects the function of the thyroid gland and interferes with the immune system. Soya contains oestrogen-like compounds, similar to those in the contraceptive pill, which can upset the body's hormonal balance. Recent research suggests that increasing consumption of soya, either directly in the diet, or indirectly in meat and milk, is diminishing male fertility and disrupting our immune systems by blocking the synthesis of thyroid hormones and interfering with the absorption of iodine essential to the proper functioning of the immune system. Western processed food contains little enough iodine without the effects of soya and all the other pollutants we are exposed to or ingest. Americans consume more than 9.3 million tons of soya bean oil every year, half of which is hydrogenated – chemically altered by adding hydrogen under heat, with a metallic catalyst such as nickel, so that it is easier to manufacture and keeps longer. Evidence is accumulating that processed soya and corn consumption are not unconnected with the epidemic of obesity and ill-health in the US. Is it a good idea to be putting this into people's bodies? The FDA may be right that such chemicalized foodstuffs do no harm to human health, but where is the benefit, other than to big business and industrial farming?

Over the last 60 years, huge industrial, 'bio-secure' dairy farms with thousands of Holsteins have almost taken over the supply of dairy produce in the US. They are closed to everyone except visitors who have made an appointment to visit. Some of the more media-savvy operations offer guided bus tours organized by their public relations people. Fair Oaks Farm in Indiana is one of the biggest and slickest of these operations. Its tour buses are painted black and white to look like the markings on a Holstein cow. As the bus progresses slowly round the farm, a recorded voice feeds the passengers the litany of bigger and better statistics that so impresses the American psyche. Fair Oaks owns 19,000 acres of land. It has 30,000 cows that live in ten barns the size of industrial units, which is what they actually are. Four hundred people look after the herds, which produce 250,000 gallons of milk a day. They boast that unlike more than half of dairy producers in the US they don't use artificial hormones to stimulate milk production. All the waste from the cows is processed through a bio-digester, which produces enough methane to generate the electricity the huge operation needs.

Milking continues twenty-four hours a day, seven days a week, with each cow being milked three times in one of the ten milking parlours. After each milking the parlour is cleaned automatically before starting the next one. Visitors can climb up to a glassed-off viewing area (bio-security again) above the milking parlour, where a huge carousel slowly rotates 72 cows at a time. It reminds me of William Harley's nineteenth-century milking operation in Glasgow, where people could buy a ticket to a viewing platform in his cow house to watch his herd being milked. The only difference is he didn't have electricity.

You can go to the 'birthing barn' to watch, from behind a mesh and glass barrier (more bio-security) the arrival of one of the 80–100 calves born every day. Female calves are destined to become replacements for their mothers in the dairy, but male calves are just a nuisance and are slaughtered as soon as possible. As do most modern intensive dairying operations, Fair Oaks inseminates the cows with sex-selected semen, ensuring that 80 per cent of calves will be female. These are then sent to be reared in Kentucky, Tennessee and Missouri on pasture farms. There they are artificially inseminated, and when seven months pregnant, at two and a half years old, they are brought back to Fair Oaks to one of the huge barns, where they will stay as long as they can produce enough milk to justify the cost of keeping them – on average three to four years. They will never again step outside the barn until the day they are sent for slaughter.

Fair Oaks is not only a ruthlessly efficient industrial farming operation, it is a slick public relations set-up as well. They have disingenuously adopted the language of the small local farmers who are trying to carve out a market directly with the public by stressing that their produce hasn't travelled very far, is from 'farm to fork', kind to the environment, sustainable, eco-friendly and so on.

There is so much to take issue with here that it's hard to know where to start. The farm is an industrial facility that uses huge amounts of energy. Every scrap of feed and all the waste from the cows has to be carried to and from them. We get a glimpse of the amount of diesel used when Fair Oaks boasts of *saving* two million gallons by extracting the methane from cow and pig muck. But putting the waste through digesters means that the huge amounts of energy the farm consumes are obtained from a different source – namely the fertility that would have gone back on the land. There is no mention of the thousands of tons of nitrogenous fertilizer, extracted from oil, they use to replace that lost through the digestion process. No mention either of what they grow to feed the cows – how much GM seed and glyphosate is used on the soya and maize.

The ultra-high-yielding Holstein cows used in this kind of farming are giving enormous amounts of milk, but are worn out by the age of six, after three to four lactations. There is therefore much greater cost in breeding replacements because between a fifth and a sixth of the herd has to be replaced every year, compared with a pasture-fed lower-yielding breed of dairy cow, which routinely achieves eight, ten or more lactations, depending on breed.

High-yielding herds cannot afford to have cows that will not get in calf. Therefore there is a cull of about 12 per cent of 'barren' cows in such herds, compared with 5 per cent or fewer in pasture-fed herds. Pasture-fed cows produce less than half the milk – about 3,000 litres a year – of the Holstein industrial herds, which average 7,600 litres in a lactation. In a single year, this is about ten times the cow's bodyweight.

Although, to its credit, Fair Oaks does not give its cows hormones, a good deal of the high output from American dairy herds is achieved by administering a synthetic version of bovine somatotropin, or somatotrophin (abbreviated bST or BST), which is a hormone produced in the pituitary gland of cattle. About half the dairy cows in the US are routinely injected with the substance. The biotech company Genentech discovered and patented the gene for the hormone in the 1970s, and allowed it to be made artificially. This gave the American scientific establishment the excuse to invent yet another acronym, rBST, which stands for recombinant bovine somatotropin. Monsanto was the first to obtain FDA approval for rBST, which they sold as Posilac. They disposed of their interest in it in 2008 to another big pharma company, Eli Lilly and Co.

The hormone is banned in Canada, Japan, Australia, New Zealand, Israel, Argentina and the EU, yet the FDA says it is safe for human consumption. It may be harmless to humans, but it is far from clear that it is harmless to cows. An EU animal welfare report concluded that it can cause 'severe and unnecessary pain, suffering and distress' to cows, from serious mastitis, foot disorders and reproductive problems. It is also doubtful whether its use increases profit by much, if at all. When the cost of extra feed, veterinary bills to deal with mastitis and other ailments, and the shorter life span of cows are taken into account, the extra 10 to 15 per cent of milk is hardly worth its use.

Compared with the US, where they have had mega-dairies since the 1960s, the UK lags far behind. One of the pioneers in England of the expansionist course is David Metcalfe, from Leyburn in North Yorkshire. When I met him at Washfold Farm in 2016, he was milking 900 British Holsteins, producing 22,000 litres (5,000 gallons a day), all sold wholesale to Paynes Dairies at Northallerton. He was in the process of increasing the herd to 1,300 cows by building a vast new shed and milking parlour at a cost he wouldn't disclose, though it's bound to be into the millions. The new farm buildings are stretched out like units on an industrial estate, connected by concrete roads, dwarfing the original farmhouse and stone buildings. The enterprise has expanded beyond anything imaginable 70 years ago, when his grandfather started farming there with 18 Dairy Shorthorn cows.

Washfold's pedigree British Holstein herd averages 10,800 litres (2,375 gallons) per cow per year, at 3.8 per cent butterfat and 3.2 per cent protein. If the average length of lactation is 320 days, each cow is averaging nearly seven and a half gallons of milk a day. This is a phenomenal yield. They show the best animals in the herd and regularly win prizes for their quality; in 2012, they were awarded the prize for the best herd in the North Eastern Holstein Club competition. Everything is highly efficient and competitive. They use sex-selected semen from the best AI bulls. They also sell pedigree bulls.

This is serious industrial dairy farming, with everything pushed to its full capacity. In the interests of efficiency, the cows never leave the sheds and are milked three times a day. Everything they would graze is cut in the fields and carried to them, and all their muck is taken away mechanically. Their daily forage ration is scientifically calculated and made up of winter wheat harvested as 'whole crop' when it's green, plus grass and whatever protein supplement is added to make their 'TMR' – an American acronym for total mixed ration – which contains the bulky part of what the cows need for maximum production. The rest comes from high-protein cow cake fed in the milking parlour according to yield.

David Metcalfe wouldn't tell me how much money he was losing on every litre of milk the farm produces, but he says he can afford it for now. His heavy road haulage business subsidizes the farming, and his bank is happy. The only thing on the farm that turns a profit is the bio-digester extracting methane from cow muck and powering an engine to generate electricity.

'It's coming to something,' I said, 'when the slurry from 900 cows pays better than selling their milk.' He didn't reply, but made a gesture that meant something like 'that's the way it is, we've got to live with it'.

He says he is 'hanging on by his fingernails' in the hope that in the next couple of years most of the farmers now contributing to the British and European milk surplus will go out of business. There will be a shortage, the price will rise dramatically, and these mega-producers will have a monopoly and 'clean up', as he puts it. That's the theory anyway.

Until the 1980s, most of Britain's milk was supplied by thousands of small farms spread across the country. But if current trends continue, the future of milk production will be the way the Metcalfes do it: in huge operations with the financial clout to take on the dairy processing companies and the supermarkets. In the US, 85 per cent of family dairy farms have disappeared in the last 40 years and the number of mega-farms with more than 2,000 cows has increased by over 100 per cent. What happens in America usually happens here sooner or later.

As we drove around on a fine, still morning, I noticed some of the cows gazing out over the metal gates across huge grass fields that stretch away from their massive building and wondered aloud if they might not be longing to be out in the pastures, grazing in the spring sunshine. 'They've never eaten grass so they don't know what they're missing,' he replied. I don't think I could have resisted letting them out to graze, even if only to relieve the monotony of their lives and see what they would do. But that would have ruined the whole tightly controlled system. And actually the cows aren't suffering. They have everything they could want – except sunshine on their backs, grass long enough to wrap their tongues around and pull, and the daily exercise of walking to and from their pastures.

But they are cows, and so far as we know, they are not given to introspection or existential angst. Moreover, they're *dairy* cows, a highly unnatural kind of bovine bred to produce milk, and lots of it, much more milk than a whole tribe of calves could ever consume. Whether they were inside or outside they would still exist to produce milk for us. It seems to me that there is no reason to feel sorry for them being confined to sheds all their lives, with everything laid on. I can't see much difference between their existence and that of the average coal-miner – or, in our own times, Amazon warehouse worker – who *are* given to introspection and angst.

It seems that a more realistic and less sentimental way would be to see it as a further example of the intimate and eternal relationship we have with cows, and of how much we rely on them. Spending their lives in huge sheds is no worse than being tied by the neck in a dark byre all winter, only able to stand up and sit down. The more important difference is that cows that would otherwise have been dispersed amongst a hundred farming families in a hundred villages are gathered together in one place. Might a more pertinent criticism of this kind of intensive dairy farming be to ask where it is all leading and whether it is right that farming is being concentrated in fewer and fewer hands. And is the milk they produce as good as that from traditional cattle grazing grass?

The leggy, large-framed Holstein has come to dominate industrial dairying in Western countries. But it is a high-maintenance beast, not bred to survive on grazing alone. It is definitely not a dual-purpose cow – its carcase is worth very little at the end of its short life – so it has to justify the cost of its keep by producing large amounts of milk. It is very much a cow for our modern throwaway world: bigger, faster and disposable, just like the consumer goods that we produce with shorter and shorter lives. It even breeds like the modern Western family, having fewer than three calves. It is an animal treated as an object, which must fit in to a vast industrial dairying operation and produce like some bovine Stakhanovite during its short existence on concrete.

But cows produce their highest yields during their fifth, sixth and seventh lactations. And the best cows from the more traditional breeds will go on to have ten, twelve or more lactations and produce more milk in total over that time at a much lower cost in terms both of rearing replacement cows and being able to produce from home-produced feed. Unlike the Holstein, their calves will find a ready market either for rearing for beef or for crossing with beef bulls to breed suckler cows for beef production.

Is it to our national benefit that dairy farmers should be forced into beggar-thy-neighbour capitalism, and that within a few years only a few massive industrial farms will be producing all our milk? They will be easier for the state to control, of course, which will suit those who see farmers as bloody-minded individualists, who despoil the land, but will it benefit our society to denude the land of the people who live on and work it? Where will the country people come from who know their own land and maintain the drains and hedges and walls? And how will the health of the soil react to this industrial onslaught?

CHAPTER 7

The Miracles of AI and Pasteurization

TWO INNOVATIONS HAVE transformed dairying in the last hundred years: artificial insemination (AI) and pasteurization. AI is almost a miracle, and probably the innovation that has had the most profound effect on cattle breeding in history. It has transformed the quality of stock, allowing farmers to select semen from the best sires in the world, from a catalogue, without needing to own a bull, and storing it until they need it.

As with many innovations in animal and plant breeding, AI began with a Dutchman, Antony van Leeuwenhoek (1632–1723), a man of many parts, none of them scientifically trained. He was an amateur whose thirst for knowledge led him to discoveries that eluded the conventionally minded establishment. Fascinated by natural history, he trained himself to be an expert grinder of magnifying lenses, the better to observe minutely the things that interested him. Through his lenses he studied the bacteria in his mouth that caused tooth plaque, the minute organisms in water and the structure of blood cells. He described all these in a series of letters he wrote over many years to the Royal Society. Then in

1678 he gave an account of his observation of moving sperm, which he called 'animalcules', through a lens he had ground to 270x magnification.

This knowledge was built on by Lazzaro Spallanzani (1729–99), an Italian priest from Pavia, who in 1784 artificially inseminated a bitch that gave birth to three pups. A hundred years later, Walter Heape (1855–1929), a biologist from Cambridge, reported that AI had been successfully used in occasional experiments to breed rabbits, dogs and horses.

But serious progress was not made until the beginning of the twentieth century, when a Russian scientist, Ilia Ivanov, achieved international recognition for his work with AI, which he originally undertook to improve Russia's imperial bloodstock. He harvested sperm from the best stallions, and by 1922 could boast that with AI one superior stallion could sire 500 foals in a year, compared with 20–30 by conventional service. He also experimented with hybridization of different species of domestic animal, creating a zeedonk by crossing a zebra with a donkey; a zubron from a European bison and a cow; and trying (without success) various combinations of rats, mice, guinea pigs and rabbits. However, when his suggestion that it might be possible to cross a human with an ape was met with widespread repugnance, he knew he had gone too far for conservative Orthodox Christian Russia.

He had to wait for the Bolshevik revolution to give him his opportunity. After Lenin's death in January 1924, he petitioned the new Soviet regime to allow him a trip to Africa to collect apes for insemination. They gave him exceptional permission to travel abroad and awarded him the huge sum of $200,000 to fund his research. This was at a time when the country was in turmoil, almost bankrupt, and millions of people were starving.

On his first trip, via Paris, to Guinea, in 1926, he failed to obtain any apes and returned to Paris, where he spent some months with the celebrated Russian émigré surgeon Serge Voronov, originator of the modish monkey gland 'rejuvenation therapy', which he claimed could reinvigorate men and spectacularly prolong human life. Its main effect was to separate rich, credulous men from large amounts of their money. During the 1920s and 30s, Voronov made a fortune. He lived in Paris like a prince with his retinue, occupying the entire floor of a hotel, and was lionized for his remarkable but doubtfully effective solution to 'ageing' – a euphemism for impotence.

When Ivanov finally returned to the Soviet Union, he advertised for women who would volunteer to bear a half-man, half-ape foetus. By the time he was ready to inseminate the first of five women who had volunteered, 'in the interests of science and the motherland', his only surviving ape, a 26-year-old orang-utan called Tarzan, had suffered a brain haemorrhage. Then, in 1930, Ivanov was arrested in one of Stalin's periodic purges and exiled for five years to Kazakhstan. Released after a year, his health broken, he suffered a stroke and died soon afterwards.

But that did not end the Bolshevik regime's fascination with AI, which they believed could accelerate the spread of 'desirable traits' in the population, such as willingness to accept communal living and working, and eradicate the unfortunate human instinct to be competitive and to own property. Through AI, human nature could be changed in consonance with the Marxist plan. Crossing the Russian peasant with an ape might have been a surer way to create *Homo sovieticus* than waiting for him to accept the Bolshevik utopia.

By the time I got into farming, AI of cattle had become available almost everywhere in Britain by phoning the nearest Milk Marketing Board AI centre the day before, or early in the morning of the day you needed the cow serving. That was all very well, but catching a cow that was out in the fields, at exactly the time of ovulation, was a different matter altogether. In the days before mobile phones, you didn't know precisely when the AI man would turn up, so the cow had to be brought in early and tied up somewhere quiet to await his arrival. With skittish young heifers this could sometimes be a bit of a rodeo. Often you had to get the whole herd into the yard, separate the ovulating animal from the others and then take the rest back to the field, which could be some distance away. If you had 30 heifers to inseminate (preferably so that they would all calve within the same month, nine months later), it is not hard to see why AI might not have been the best method, and why I find myself in Chapter 10 on the road to Hereford to buy a bull.

Even back then, when I was part of that world, I thought being an AI man was a pretty odd job, which attracted a strange type of person. Rushing round the countryside, from farm to farm, to catch a cow in season and squirt deep-frozen bull semen into her vagina would not be most people's first choice of career. Although there were a few women doing it, most of the AI operators in those days were men; often mild-mannered, even shy, their ready embarrassment juxtaposed with the earthiness of their job afforded great scope for amusement.

One summer I had a 15-year-old lad, Carl, helping during the school holidays. He was from the town and wasn't too bright. He'd been advised by his careers master that farm work might suit him when he left school, the implication being that as farmers are a bit stupid, he would fit in perfectly. The AI man was coming that morning to inseminate a cow that I'd spotted in season – 'bulling' – but he still hadn't arrived by the time I had to leave to go to the auction mart. I asked Carl to keep a lookout for him. Carl had no idea what AI was and I had to explain.

'There are three black cows tied up in the byre,' I added. 'The one nearest the door is the one to be inseminated. There's a big wooden peg in the post beside her so you will know which one she is.'

I took him into the byre, showed him the cow and pointed out the peg.

'What's the peg for?' he asked.

'For the AI man to hang his trousers on.'

He nodded sagely and I left.

The next time the AI man came, he said, 'That's a strange lad you had working for you last time I was here. I've heard the old joke till I'm sick of it. But he wasn't joking. He loitered around and twice showed me where to hang my trousers. When I'd done the job he looked really surprised and asked if that was all there was to it. God alone knows what he expected me to do.'

When I told the story to a farmer friend who knew Carl's family, he almost fell off his bar stool laughing. Apparently, after the war, Carl's uncle had served a prison sentence for having carnal knowledge of a sheep.

'It must be genetic,' spluttered my drinking companion.

In the 1930s, artificial cow vaginas were invented to collect semen from bulls called 'mount animals'. And in 1936, over 1,000 cows were inseminated in Denmark, with nearly 60 per cent of them conceiving – almost as good as with natural service. The Danish professor who did much of this work, Eduard Sørensen, got the idea of using straws for the storage of semen and its deposition deep into a cow's uterus when he saw his daughter's friends at her birthday party sipping punch through oat straws.

This stimulated an avalanche of research and development in other countries, particularly in the US during the 1940s. It was crucial to success that sperm could be kept alive longer than the few minutes it remained viable outside the bull's body; and that some process could be devised to make each ejaculate go further.

Researchers found that a mixture of egg yolk and sodium citrate would 'buffer' the semen, causing it to resist changes in pH and allowing it to survive for up to three days at 5°C. Sodium citrate is used to prevent UHT milk from coagulating; when added to any cheese, it gives it the texture of processed cheese but supposedly allows it to keep the flavour of the original. This is popular in the US, where it is called constructed cheese and is based on a chemical process developed by James L. Kraft, who in 1916 created the first emulsified 'melty' cheese slice and a vast fortune for himself.

With the discovery of penicillin and streptomycin, it became possible to control certain bovine venereal diseases that tended to be passed on with AI. It was then found that adding milk to the semen 'extended' it – that is, made it go further – so that it could be used at lower concentrations and more inseminations could be obtained from one ejaculate. It was also found that caproic acid (the fatty acid that gives goat's cheese its distinctive smell) and catalase (the enzyme that protects living organisms against decomposition) added to 5 per cent egg yolk (a drug called Caprogen) would preserve semen at ambient temperatures. It also extended semen so that effective conception could be achieved with between 26^6 and 106^6 sperm per insemination.

Hardly surprisingly, researchers found that different bulls were excited by different sexual stimuli. So by catering to whatever turned him on, each bull could be manipulated into giving as many as six ejaculations a week, providing between 30^9 and 40^9 sperm per week per bull. This is about 200,000 doses of semen per bull per year. They also confirmed the long-held belief amongst cattle breeders that the bigger the testicles, the more semen is produced.

But this was all of limited application because they could not keep semen alive in storage. In 1949, an English researcher into cryo-preservation, Christopher Polge, had preserved chicken sperm at low temperature by adding fructose and had produced chicks from eggs fertilized with it. But he could not get this to work with bull semen. After leaving it for six months, Polge decided to try his sugar method again. He added what he thought was the same fructose solution from the bottle he had used earlier, but this time, to his surprise, he achieved more success. When he analysed the solution he had used, he found it contained no sugar, but glycerol and protein in the same proportions as make up Mayer's albumin, an adhesive (like egg white) used for affixing specimens to glass slides. Somebody had labelled the bottle wrongly. Thus is scientific progress made.

What Polge had in fact stumbled upon was the original egg yolk/citrate 'extender' that had first been used in 1941. Later, egg yolk buffered with the enzyme inhibitor tris and mixed with glycerol was found to be most effective at protecting sperm from harm at very low temperatures. It was discovered that bull sperm could be stored viably for a long time in solid carbon dioxide at a temperature of minus 79°C, and stored in liquid nitrogen at minus 196°C, it would survive almost indefinitely. The problem of the glass ampoules containing the sperm tending to shatter during freezing and thawing was solved by using sealed plastic straws that could be fitted into a gun with a long barrel and plunger for insemination. And when it was found that a dose of semen could be reduced to 0.25 ml per straw, so that twice the number of doses could be stored in the same space, modern AI was on the way to revolutionizing animal breeding. The final hurdle to wide-spread, affordable AI was surmounted when the American Cyanamid Company began to make insulated portable can-isters for the long-term storage of liquid nitrogen.

These developments that made possible the cryo-preserva-tion of sperm underpin the modern human fertility programme. Without them, none of the in vitro fertilization, artificial con-ception, cloning or other techniques in human and animal fertility across the world would have been possible.

Ordinary cattle breeders can buy semen from the best bulls, which would previously only have been available to the richest pedigree breeders. They can select a different bull for each cow in their herd. Such a thing could never have been possible with conventional breeding, and it has resulted in a huge improvement in quality and productivity, particularly of dairy cattle, in all the pastoral countries of the world. In a few decades, the average yield of all dairy cattle has increased beyond anything our ancestors could have imagined. And now, with sex-selected semen, it is possible to breed almost exclusively heifer calves, which suits the dairy farmer and obviates the wasteful killing of unwanted bull calves.

I had no idea that behind my herd of cows lay a web of science stretching halfway round the globe. Watching them quietly cropping the grass in my fields seemed the most natural thing in the world, the closest you could get to there being a simple relationship between farmer, animal and land. It assaulted my romantic sensibilities to know that their existence depended on such complicated scientific processes. I didn't want to know that their father's sperm had probably been harvested inside an artificial vagina in a breeding station somewhere in New Zealand, then frozen in liquid nitrogen and flown thousands of miles until it was eventually carried in a van in an insulated steel canister to a little farm, where a dairy cow was impregnated by semen from a plastic straw, shot from a gun, by a man in a brown overall.

Echoing John Maynard Keynes's aperçu that 'practical men who believe themselves to be quite exempt from any intellectual influence, are usually the slaves of some defunct economist', it is salutary to observe the hold that the dicta of Louis Pasteur have on public health policy across the Western world.

There is no doubt that in Victorian town dairies, hygiene was sometimes dreadful. Milk was sold by the ladleful from open pails. *Punch* joked that London would have to wait 'for a February with five Sundays to be able to get a clean glass of milk'. Public health campaigners began to agitate for the heat treatment of milk based on Louis Pasteur's findings that germs did not spontaneously generate, as had been thought, but grew from other germs and could be destroyed by heat. In 1893, New York was the first city to make pasteurized milk compulsory, when Nathan Strauss opened the first pasteurization plant after his daughter died from tuberculosis apparently contracted from infected milk. Pasteurization was introduced in London a few years later, as a temporary measure, mainly to stop the spread of TB, and is said to have reduced by half the deaths of babies from infantile diarrhoea.

Heat treatment is now routine in Europe and the Anglophone countries and in many is enforced by law. Pasteurization involves heating the milk and cooling it immediately afterwards, either for a short burst of 16 seconds at 72°C or a longer exposure for 30 minutes at 63°C. The higher the temperature, the more microorganisms, enzymes and vitamins are killed. Milk can also be sterilized (it is not actually sterile) by heating it to about 115°C for 20 minutes. In the process, it loses most of its vitamins and enzymes. Ultra-heat-treated (UHT) milk is flash-heated to 135°C for a second; the process kills good and bad bacteria and destroys or damages the vitamin, enzyme and other nutritional content, and it loses much of its taste in the process. In fact UHT milk is so unnatural that bacteria won't touch it. It will keep without refrigeration for up to 60 days in plastic bottles and up to five months in sterile glass ones.

In Britain, with our history of drinking fresh liquid milk, only about 8 per cent of the milk market is UHT because we don't like the tasteless stuff, but almost all of it is now pasteurized. By contrast, in hotter countries in Europe most of the milk is UHT. In 2008, in an effort to conform to European standardization, the UK government proposed that 90 per cent of our liquid milk sales should be UHT by 2020. The stated reason was to reduce the need for refrigeration and reduce greenhouse gas emissions. Mercifully, the big dairies and milk processors opposed it and the proposal was abandoned.

The target organism that had to be killed if pasteurization was to be effective was the germ that causes TB, *Mycobacterium paratuberculosis*. But testing for this took between 24 and 48 hours and the milk could have gone off before success could be verified. So another test was developed based on an enzyme called alkaline phosphatase (ALP), which is present in the milk of all mammals. It requires slightly more heat to kill it than the target organism and its presence is easily tested by measuring whether the milk becomes fluorescent when exposed to active ALP. This is now the standard method in the UK to show that milk has been pasteurized successfully. Unfortunately, destroying the phosphatase enzyme impairs the body's capacity to absorb calcium. Pasteurization also alters or destroys many amino acids, and reduces the digestibility of protein by about 17 per cent. Beneficial bacteria that prevent milk from decomposing are also destroyed, so that pasteurized milk goes rancid rather than souring.

Pasteur's theory, which affects the diet of almost everyone in the West, is that illness is caused by 'germs', which must be destroyed to maintain human health. Pasteur did not go unchallenged, particularly by his great rival, another French scientist, Antoine Béchamp, whose counter-theory based on the principle of 'wellness' animates an alternative view of health and illness.

Béchamp made the point that our bodies contain a colony of over 10,000 interdependent species and subspecies of bacteria, particularly in the gut, without which we would not survive for long. This vast and complicated range of 'good' and 'bad' bacteria (Pasteur's 'germs') works like the ecosystem in a rainforest and maintains a balance in the human body; illness only strikes when something upsets the equilibrium. Béchamp called this 'cellular theory' or 'vitalism' or 'microzymian' theory. A healthy body is protected against illness unless something goes wrong to cause it. Treating illness therefore involves restoring the basic conditions that promote health and allowing the body to bring itself back into order.

This contrasts with Pasteur's view that disease defines life negatively, rather than as a positive force. Mainstream Western medicine largely accepts Pasteur's model, and takes it for granted that pasteurization is necessary to maintain the health of the population. Antibacterial handwashes, kitchen and bathroom cleaning fluids and disinfectants that kill 'germs' and preserve us from disease find a ready and highly profitable market in the West.

Béchamp and Pasteur had a long and often bitter professional rivalry. But in the end, Pasteur's ideas found a readier reception than those of his more sophisticated, subtle and perhaps more challenging colleague. Pasteur's theory informs all public health legislation across the Western world. Diseases are caused by particular microorganisms (germs) that invade the body and make it ill. They can attack anybody at any time, irrespective of whether or not the individual takes care of himself. This largely absolves the sufferer of responsibility for his own health.

Béchamp did not deny that microorganisms can cause disease, but he believed that in most cases they do not invade from the outside; rather they are present throughout the cells of the body and both maintain its life (metabolic) and aid in its disintegration (catabolic) if it is injured or dies. Microorganisms are not immutable, but are living things that adapt to the conditions they encounter. Every disease arises because the underlying health of the body has been compromised in some way. People can reduce their susceptibility to disease by living and eating well and by listening to their bodies, which will remain healthy if treated with respect and good sense. Most alternative medicine is founded on these principles. And actually, much of human health depends on them. Doctors do not maintain good health. They can prolong life, but often it is by creating the conditions for the body to heal itself.

In 1919, Professor Henry Armstrong observed in an article for the *Journal of the Royal Society of Arts* that it was 'astounding the extent to which Pasteur had influenced our doings', and remarked that 'the real harm was done when milk was tampered with', because its dietetic value was diminished by heating it anywhere above blood temperature. Sterilizing milk undoubtedly reduced the risk of infection from typhus and tuberculosis, but destroying the 'lactic organism' encouraged the growth of 'putrefactive organisms', causing cases of infantile diarrhoea and lactic intolerance.

It is widely recognized, and not only by its opponents, that pasteurization damages milk and makes it harder to digest. In 1932, the physician to the royal family, Lord Dawson of Penn, warned that 'pasteurization could never make bad milk into good milk'. He argued that a better solution to the problem of disease would be to clean up dairies and improve hygiene on the farm rather than damage a highly important part of the national diet. He was far from alone in opposing routine pasteurization. The late Queen Mother is said to have insisted on raw milk from the royal herd of Ayrshire cows at Windsor.

Considering the opposition to it, it is all the more remarkable that Pasteur's germ theory has become accepted as necessary to make milk and dairy products 'safe'. The medical establishment and public health authorities are so convinced of their absolute rectitude that in certain countries and states of the US, compulsory pasteurization is backed by the full force of law. It is not a matter over which people are allowed to make up their own minds.

Pasteur's biographer has an unattributed account of him recanting on his deathbed: 'Bernard is correct. The bacteria are nothing. The soil [*terrain*] is everything.'* He is said to have been referring to the conclusions of Claude Bernard (1813–78), the third of the triumvirate of nineteenth-century French scientists whose research into fermentation, microbes and contagious disease overlapped. Bernard's theory was similar to Béchamp's and described the '*milieu intérieur*' by which the cells of the body maintain it in a state of equilibrium. The medium in which bacteria grow is the defining thing, and if the germs do not find a fertile soil for their growth, they will have no effect. 'It's not the mosquito, but the swamp it grows in.' Pasteur collaborated with Bernard, but went to some lengths to denigrate Béchamp's work, which threatened

---

* Ethel Douglas Hume, *Béchamp or Pasteur? A Lost Chapter in the History of Biology* (Chicago: Covici-McGee, 1923; London: C. W. Daniel Co., Ltd).

his reputation and income. He has been accused of not being afraid to falsify the results of his experiments if it suited his purpose and to denigrate the theories of the more self-effacing Béchamp.* As a result, Béchamp is almost unrecognized today, even though his work supports the theory that underlies the practice of much alternative medicine.

As well as being pasteurized, most of our milk is now homogenized. This process was invented by another Frenchman, Auguste Gaulin, to prevent unscrupulous dairymen from skimming off some of the cream and passing off what was left as whole milk. The milk is forced, under great pressure, through tiny tapered tubes, which breaks up the fat globules into smaller particles and mixes them with protein particles to form a solution. This creates a uniform product with a creamy consistency that doesn't need shaking up before the container is opened. When it was first introduced to the US in 1919, customers were suspicious of such messing about with a natural product and thought (with some reason) that it would make it hard to digest. One dairy in Michigan got its employees to drink the product in front of sceptical housewives and then vomit up the partially digested milk curds, to try to convince them that homogenized milk was easily digestible.

After the fat has been broken up, the smaller globules can attract fragments of whey and casein and sometimes are completely surrounded by a layer of protein. These chemically altered globules of fat and protein tend to clump together and must be put through the process again to break them up

* Marie Nonclercq, a French pharmacist, wrote for her doctoral dissertation a biography of Béchamp. It was published as a book in 1992 by Maloine as *Antoine Béchamp, 1816–1908: The Man and the Scientist, the Originality and Productivity of His Work.*

further to make a permanent solution. The milk is usually pasteurized first, to kill the enzyme lipase, which would otherwise start to digest the ruptured fat globules and turn it rancid. The high pressure of the homogenization process generates enough heat to pasteurize the milk a second time.

Pasteurizing milk benefits the processor and does not necessarily provide a superior fresh product. The average pint from a supermarket has seen a fair bit of the world before it reaches the customer's breakfast cereal. Its journey begins with the cows being milked and the milk passed into a refrigerated bulk tank on the farm, where it remains for up to 24 hours – sometimes two days. It is then collected by tanker and carried to the processing dairy, which can be a hundred or more miles away, where it is mixed up with other milk and kept until it is pasteurized – sometimes for as long as four days.

But it is the next process that most people are unaware of. The product is 'standardized' by separating all the cream from the milk and then adding back however much cream is needed to make it either 'whole milk', 'semi-skimmed' or 'skimmed'. Any cream left over goes into other products. Even what is sold as whole milk does not always have all the cream added back. The milk with cream added back is then homogenized, packaged ready for distribution and carried to wherever the processor has customers. It is not unknown for milk to travel hundreds of miles before ending up at a supermarket in the town next to the farm it came from. Because pasteurization prolongs its life significantly, the milk can be anything up to two weeks old by the time it reaches the customer's fridge.

CHAPTER 8

Raw Milk

SHORTLY AFTER I started milking cows, I took over a round in the village from a neighbouring farmer who had decided to retire. He had delivered about 150 pints of unpasteurized raw milk ('green top', from the colour of the cap) every day for decades. He also sold home-grown vegetables: potatoes, cabbage, carrots and swedes – or turnips, as we called them. My father grew most of the vegetables we ate at home, but occasionally we ran out of something and had to buy it. I remember my mother asked our farmer neighbour for a turnip, which he delivered next day. For some reason I could not understand at the time, she was furious at the price he charged her for it. 'Fancy charging that for a turnip, the greedy little man!' I thought it was perfectly reasonable, but she wouldn't let the subject drop. 'It's not as if he hasn't got plenty. There's a field full up there!' She never really forgave him for being 'tight' and ever afterwards used his name as a metaphor for meanness.

I had never bottled or delivered milk before I took over his round. But after going with him once at the end of the month, the next day he handed over his round book, a few crates of empty bottles and the ageing, well-used bits of equipment he used to do the bottling, and next morning I was on my own. He gave me the goodwill of his round, and his equipment, thus confounding my mother's belief that he was tight.

Every morning after that, once the milk had cooled in the tank, I filled and capped 150 wide-necked pint bottles ready to be loaded onto the back of my pickup after breakfast. I would then set off with Spot, my hairy collie, in the passenger seat. Sometimes one of the children came as well. It was usually Libby, who was about two at the time, and struggled to carry four pints of milk in a little wire carrying crate. She could see through the windscreen if she stood up in the footwell and pressed her palms on the dashboard, bossing the dog about if he got in her way or wasn't sitting right, and shouting, 'Phot!' because she couldn't say her 'S's. Spot was immune to her rebukes, ignoring every command except mine – and on occasion, mine as well.

If I got a move on, it took about an hour and a half to do the round. I had it down to a fine art and amused myself by trying to find ways of doing it quicker each time. Part of the main street in the village was on a slight slope and had a high flagstone kerb along the edge of the pavement. I had one of those elastic bungee cords with a hook on each end that people use to tie luggage onto their roof racks. If I fastened a hook onto one spoke of the steering wheel and stretched it tightly round the driver's seat and back onto the other spoke, then set the pickup off in bottom gear, with the engine just ticking over, it would more or less drive itself in a straight line down the road, occasionally bumping against the kerb, which corrected its course. I ran from house to house with the milk crate, in and out of the gardens, just about keeping up with the pickup, now and again leaning in through the window and tweaking the steering wheel if it was going off course.

Mercifully there was little traffic in those days and I knew everybody who was likely to come up the road. Using this method I could do the work of two people – driver and delivery boy. I could only do this on days when I didn't deliver eggs or collect the money. If anybody increased their order – as they did regularly by leaving a note on the doorstep – I had to run a bit faster to catch up with the pickup as it ticked along down the road, grab the extra bottles and run back into their garden.

Saturday was the best day of the week, when I would usually go home with not less than £150 in cash – a great deal of money in the early 1980s. My monthly milk cheque from the MMB came to about £3,000 for around 250 gallons a day (1,100 litres) – the milk from 60 cows – whereas for selling 14 gallons (60 litres) of milk a day in bottles – the milk from three cows – I got £600 a month. I think the Milk Marketing Board paid something like 8 or 9p a litre, whereas the statutory retail price for a pint was around 20p – roughly 35p a litre.

Some people in the village refused to take my milk because it was unpasteurized, straight from the cow. Others bought quantities *because* it was unpasteurized. One customer, who had retired to the village from some high-flying civil service job on a huge pension, credited my milk with life-saving properties. He had developed ulcerative colitis and his doctor told him he would need a colostomy to save his life. Just before the operation, he consulted an alternative healer, who told him that yoghurt made with unpasteurized milk might cure his bowel problems. And it did. He made an almost complete recovery by eating a pint of yoghurt a day made with my milk. At the time I took what he said with a pinch of salt, because I thought he was exaggerating the benefits of raw milk. But over the years I've heard too many stories about its healing properties not to take them seriously – it can cure chronic digestive problems and skin ailments, and some people even claim it shrinks tumours.

Not every one of my customers was equally delighted with natural milk. One Saturday morning when I knocked for the week's money, one of my more hygiene-conscious patrons opened the door grasping a full bottle by the neck and lifted it up to point to an unmistakable ring of grey-brown sludge at the bottom that disappeared when she shook it up. She claimed all the milk I had delivered the previous day had been similarly tainted. I was rather worried about this because at the time the public health people were just getting into their stride looking for reasons to stop farmers selling untreated milk. Despite their best efforts they had never been able to find anything wrong with mine, but this might have been the pretext they were looking for.

My customer was unmoved by my offer not to charge her for that week's deliveries, so rashly I offered not to charge her for *any* milk for the rest of the year if she got so much as one more bottle with sludge in it. With rather bad grace she accepted what, in anybody's view, was a tremendous offer. It wasn't as if her family ran any risk from consuming my milk: all my other customers and my own family drank it without ill effects. It just had fine bits of mineral matter in it, rather like Morbier cheese with its layer of wood ash through the middle, or a decent bottle of claret that throws a sediment. I didn't dare make these analogies, because she had a kitchen like an operating theatre and was terrified of Pasteur's 'germs'. She didn't want her milk natural, or for nothing; she wanted it pure and white and not to be reminded where it came from. Another bottle with grit in it would have caused her to cancel her order altogether.

I was a little shaken by this encounter, and thinking about it on the way home, it dawned on me how the grit had got into the milk. We had had a few wet days and some of the cows' udders had been particularly muddy when they came in for milking. Unless they had cow muck on them, I didn't normally wash their teats, because washing udders tends to spread mastitis, which is the bane of a dairyman's life. As a result, my herd's mastitis cell count – which the dairy measured and showed on my monthly statement – was very low. I simply rubbed off any dry soil and relied on a filter called a milk sock, which fitted on the end of the pipe that took the milk into the bulk tank and which I changed every milking. I remembered that at some point during the previous evening's milking the sock had blown off the pipe. I replaced it with a fresh one and thought nothing more about it. It had never happened before and so I didn't realize the consequences. My customer never found any more grit in her milk and I had learned a valuable lesson.

Cream rises to the top because whole milk is an emulsion, a mixture of fat, protein solids and water that settles out. Homogenizing milk turns it into a colloid – a mixture of microscopic particles of milk solids dispersed in a liquid – which doesn't settle out.

Raw milk contains a full complement of vitamins B and C, but these are largely destroyed by heat treatment and they also decay as the milk ages. After about 7–10 days, most will have disappeared. Raw milk is also a tremendous source of fat-soluble digestible calcium, but if the fat is damaged by high-temperature pasteurization, its structure is altered and it becomes insoluble, so that it passes through the human body without it being absorbed. To be available, the soluble calcium must be present with soluble vitamin D so that the milk fat can dissolve them. To get the full benefit of all the calcium and vitamin D, the milk must be consumed whole and raw with none of its fat removed. Skimmed milk contains minimal fat and barely any vitamin D, and so most of the calcium in it cannot be absorbed by the body.

Fresh raw milk also contains large amounts of fat-soluble vitamin A. But if the fat in the milk is damaged by pumping or heating or removed by skimming, almost none of that is available either. Damage to the fat can happen when milk is being transported by tanker, or pumped into and out of the dairy and along pipes to the processing plant. The agitation causes aeration, leading to oxidation of the fats, which significantly reduces the vitamin content. Homogenization also has a similar effect in deforming the fat. The less milk is aerated, pumped or otherwise messed about with, the more fat-soluble vitamins are conserved. Most damaging of all is heat treatment, which also destroys most of the natural enzymes.

Raw milk is one of the most easily assimilated sources of calcium, of which there is a severe shortage in the modern Western diet, especially amongst younger people, who are afraid that milk is bad for them. They might have a point about pasteurized milk because as the dead bacteria decompose, they release histamine, which might cause eczema and other allergies. Raw milk, on the other hand, is a living food that seems to have a beneficial effect on allergies, eczema and hay fever, particularly in children. There are many reported instances of eczema disappearing within a short time in people who take up drinking raw milk.

Even when the fat-clogs-your-arteries hysteria was at its height and the margarine manufacturers, aided by tame scientists, had terrified millions of people in the US and Britain into giving up animal fats, I could see no logic to it. Granted, there seemed to be an epidemic of heart disease in America. But I couldn't see how it could be caused by eating foodstuffs that had sustained us for millennia. I thought there must be something else at play, although I didn't know what it was. It was hard to avoid the blanket condemnation of animal fats. People even became afraid to eat their breakfast egg because it contained poisonous cholesterol. I remember being advised in apocalyptic tones by the government to avoid eggs, and if we couldn't give them up completely, to eat only one a week.

The scientific research (much of it paid for by US margarine manufacturers) was so convincing and the relentless advertising campaigns so successful that it became embedded in the Western psyche that eating fat would make you fat and kill you. It inspired the dairy industry's 'Naughty but Nice' advert, an attempt to keep up the sales of cream, not by challenging the 'science' (which was publicly unchallengeable at the time) but by encouraging people to indulge themselves in a harmful but pleasurable vice – and then no doubt feel guilty about it afterwards. They were the same tactics the drink and tobacco industries used. By implication, resisting the urge to eat animal fats and dairy produce and turning to man-made products like margarine would be virtuous.

Although this nonsense has largely been discredited, it still nags away at millions of otherwise sensible people, particularly overweight ones, who are induced to buy processed 'low-fat' food products, made palatable by the addition of sugar. This attitude has affected the liquid milk market, which paradoxically has benefited hugely from the campaign against its products. The dairy processor can now get as much for milk denuded completely of its fat (red top) – little more than white water – as he can for proper milk. Skimmed milk used to be fed to pigs; now the processors can sell the skimmed milk separately from the cream and butter and cut out the pig.

One day I happened to mention the subject to my farming cousin John, a strong, lean and very fit man in his late fifties who loved fat of all kinds, particularly cream, home-made salted butter, the crisp fat off beef and lamb, and pork crackling. His particular favourite was pig-foot pie, which his wife made him as a treat – pig's trotters swimming in fat, with carrots and onions, under thick pastry, with a patty pan holding up the middle.

He had thought about the cholesterol question, as had most people at the time, because of the relentless badgering in the media from government scientists, 'health professionals' and food manufacturers. It was so pervasive and convincing that I had begun to worry that my father (who was also in his late fifties and very fit) might be silently developing heart disease because he ate animal fats, in particular butter and cream. John cocked his head to one side with one of his wry half-smiles.

'When I was growing up in the 1920s and 30s,' he said, 'we lived off the fat of the land: butter, cream, milk, eggs, as much meat as we wanted. We always had home-cured bacon and ham, black pudding and brawn, and liver and kidney when we killed a sheep. We had cock chickens, a few ducks, and a goose at Christmas; and we grew all the vegetables we needed. Every Friday my mother bought fish from the fish van. When rationing came with the war, we hardly noticed. We never went short. Nobody checked up on how much we were eating. We just carried on as before. And we helped out anybody who was living on rations if we trusted them to keep quiet. If eating fat clogs your arteries, then I must be a walking medical miracle. I think it's a lot of nonsense. The best rule you can follow is to eat as close to the soil as possible; the less your food's messed about with, the better.'

Once upon a time, my little milk round would have been nothing out of the ordinary. Every town and village up and down the land would have had a farmer or two delivering untreated milk door to door that he had produced from his own cows. Now there are only about a hundred farmers left in England and Wales who deliver their own raw milk.

Phil and Steve Hook (P. G. T. Hook and Son) from Longleys Farm, Hailsham in Sussex, cleverly thrive by producing raw milk as it used to be. They have not increased the size of their 70-cow herd or their 150-acre farm, or installed a fancy milking parlour, nor do they feed imported high-protein cattle cake, or sell their milk to any of the wholesalers, or process it in any way. They employ a staff of 17 and now claim to produce half of all the raw milk sold in England and Wales. They deal direct with their customers and charge a proper price that guarantees a decent return. They have milk rounds in Sussex on Mondays and Fridays and also send milk and butter nationwide by courier in polystyrene-lined insulated boxes. Their local deliveries are in returnable glass bottles at £1 a pint, while their national deliveries are, of necessity, in plastic bottles, but they will collect the polystyrene boxes for recycling.

Despite considerable pressure from the government to force all milk to be pasteurized, 2 per cent of consumers still want unpasteurized milk. This demand keeps the government from banning it altogether, even though it has been trying since 1946, when compulsory pasteurization was first proposed. Selling it retail in shops was banned in 1985. It can only be sold directly to consumers by registered milk producers at the farm gate (or through a farm catering operation), at farmers' markets (by the producer), or by a distributor from a vehicle used as 'shop premises' (milk rounds).

The cows must be healthy and free from brucellosis and tuberculosis, and must pass frequent and exacting health checks. The milking and dairy premises must comply with Food Standards Agency hygiene rules and, crucially, the milk must bear a cigarette-packet type of approved health warning. Raw cream does not have to have the health warning but must have 'made with raw milk' displayed on the packaging. Selling raw milk is banned in Scotland, yet the Hooks and a few other producers send it there by courier, even to the Highlands and Islands. This is an encouraging interpretation of the rules.

There is strong and growing opposition to the routine compulsory pasteurization of milk. It is arguably no longer necessary because all cows are TB tested and any found to have the disease are compulsorily slaughtered. It has been found that countries with the highest milk consumption also have higher rates of osteoporosis. This is hard to understand until you realize that in these countries it is processed, pasteurized and homogenized milk that is consumed. Processing by heat treating makes milk acidic and causes leaching of alkali from the body. There is some medical opinion that removing the good bacteria in cow's milk also affects resistance to disease and causes allergies and illness.

Many of the Hooks' customers claim their milk has cured asthma, eczema and hay fever, and lowered their cholesterol. One customer in the Midlands has it delivered every week for her son, who has cystic fibrosis, on the advice of the boy's consultant clinician. There are claims that raw cow's milk can shrink tumours. The Food Standards Agency concedes there has not been a single incidence of food poisoning linked to raw milk in England or Wales for over a decade, but it does not look favourably upon it. Nor do the big dairy processing companies, which hardly lose an opportunity to denigrate its safety. But the Hooks' milk is more stringently tested than most milk destined for pasteurization, and they are scrupulous about doing weekly laboratory tests for a range of different pathogens to make sure they keep ahead of the FSA.

From the producer's point of view, once it has been pasteurized, milk becomes a commodity and is subject to the laws of supply and demand. Raw milk, on the other hand, is a niche product for which people willingly pay a premium. Although their cows only produce about 4,500 litres a year each, their milk sells for on average eight times the wholesale price. This is what people are prepared to pay for 'white gold'. And yet in the hands of the processors and supermarkets it becomes 'white water' and is sold at a loss. The answer for dairy farmers is surely to go direct to their market and cut out the host of middlemen who profit from their product. They should take back control. This is a much more sensible and sustainable way to farm than making farms ever larger, more intensive and industrial. It also keeps families farming a little piece of their own land in the country.

If dairying had not been taken over by the big processors, we might still have locally produced grass-fed milk in every part of Britain. Not milk from cows kept in huge sheds and fed on imported soya meal and maize. But almost compulsory pasteurization means that whatever the quality of the milk, it can be made safe by heat treatment, giving the advantage to industrial dairying, and to processors who dictate the price of the commodity.

In common with 20 other US states, it is actually unlawful in Indiana to sell unpasteurized cow's milk and milk products for human consumption. In neighbouring Illinois and Wisconsin, producers may sell it from their own farms, and in Missouri they can sell it from a milk round. The US Food and Drug Administration bans the sale or distribution of cow's milk and milk products across state boundaries unless they are pasteurized and meet the standards of the US Pasteurized Milk Ordinance. As of April 2016, the sale of raw milk in shops was lawful in 13 states, and 17 permit raw milk sales from the farm where it is produced. Eight states that prohibit milk sales allow raw milk to be obtained through the legal fiction of 'cow-share' agreements, in which a person can buy a share in a cow and thus be entitled to consume his own cow's milk. It is not unlawful in the US for anyone to *consume* raw milk, though anyhow, preventing it would be impossible to enforce. And except in Michigan, which forbids the sale of raw milk for any purpose, all states allow it to be sold for animal feed.

This allows the Yegerlehner family, who farm near Clay City, to sell all the milk they can produce from their little herd of 30 grass-fed cows: they make cheese, butter, ice cream, yoghurt, in fact anything that can be produced from cow's milk. But everything they sell must be labelled 'Pet Food Not for Human Consumption'. The FDA knows full well that most of it is consumed by human beings, but there is nothing they can do to prevent people eating pet food if they want. It is an odd state of affairs that pets in America eat better than their owners.

The irony is that because the milk is sold for animal consumption, there is no proper licensing system. There are minimal, if any, hygiene inspections, and so long as producers do not make a great noise about it, the FDA turns a blind eye. In fact, Alan Yegerlehner, who makes the cheese and butter, maintains his own high standards of hygiene and has never had any complaint or reported illness from any of the pets or their owners consuming his produce. The farm is in the middle of nowhere, just like much of rural America; the nearest city, Indianapolis, is over an hour's drive away. The Yegerlehners would benefit from internet advertising for their produce, but as it's illegal, they simply cannot say publicly that it is beneficial for human health.

Their little herd is managed with the lowest possible inputs. They have gone back far beyond the modern agro-industrial complex to an earlier time when farmers sold what their land could produce without artificially altering their environment with chemicals, fertilizers and expensive machinery. The pastures are natural grassland, mob-stocked – what we would call rotationally grazed – and managed to allow the build-up of forage during the growing season so that the cattle have enough to last them through most of the winter. Their calving and milking is naturally timed to coincide with the growth of grass in the spring. And as the growth and quality of the grazing declines in the autumn, the milk yield declines with it, until by December the herd is dry and milking ceases until the first cow calves the following spring.

'We enjoy our fireside and read books in the winter,' says Kate Yegerlehner. 'It's lovely to have some time off when you don't have to get up to milk cows.' And even when the cows are at their peak production, they only milk once a day. Milking is a leisurely affair, which does not dominate the day as it does with herds that need to be forced to produce as much as possible. The two cheese rooms are refrigerated wagon trailers fitted out with shelves, and the raw-milk pet food is sealed into plastic bags to await delivery to drop-off points and farmers' markets in Indianapolis.

In some countries where pasteurization is rigorously enforced and sales of raw milk have been banned completely, the legal fiction of cow-sharing has attracted the heavy hand of the state. In Canada, in 2010, such a cow-sharing arrangement landed Michael Schmidt, an Ontario dairy farmer and advocate of raw milk, with a prison sentence. At first the magistrate acquitted him of 19 charges of distributing unpasteurized milk. But the Ontario Court of Justice allowed an appeal by the prosecution and reversed that decision. Schmidt was then convicted of 13 charges of breaching the ban on selling raw milk, fined $9,150 and put on probation for a year. His appeal to the Ontario Court of Appeal was dismissed. The court issued an injunction preventing him from distributing raw milk in the state. In 2011, he began a five-week hunger strike in protest against his treatment. Then in 2013, he was found guilty of contempt of court for breaching the injunction and sentenced to three months' imprisonment, suspended for a year. His appeal was dismissed in 2015. The story of his 25-year fight is the subject of the film *Milk War*.

Schmidt is an outspoken advocate for our right to eat what we want, and refused to keep his head down. The authorities have raided his farm numerous times, seized and destroyed equipment, and taken various enforcement proceedings against him. He is the most prominent of those campaigning against the blanket ban on raw milk being sold in Canada. The farm, which has been at the centre of the Canadian raw milk battle for over 23 years, is owned by 150 families, who are members of the Our Farm Our Food cooperative. In March 2017, the campaign culminated in a ten-day trial of four of the owners, who were charged with obstructing a police officer during a well-publicized raid and dramatic stand-off in October 2015. At the trial, the police put on a great show of force. Four heavily armed uniformed officers were in court, with three others waiting outside. The trial was adjourned several times; eventually, after nearly two years, the court found Schmidt guilty of obstructing a 'peace' officer and sentenced him to 60 days in prison, which he can serve at weekends. At the time of writing, the judgement and sentence are under appeal, without much hope of it being successful.

The police had wrongly claimed (deliberately, the defendants said) that the men were members of the Freemen of the Land, a libertarian organization classed as an 'extremist' group in Canada. Categorizing these farmers as extremists caused no end of trouble for them and their families, with their names being added to a national database of people who posed a violent threat to the state. Border agencies and other state enforcers were alerted to their dangerous proclivities, which affected their lives in a host of different ways.

The Canadian state was determined to make an example of the people involved. Farmers may have led revolutions in the past, but surely consuming unpasteurized milk from their own cows didn't require armed police to raid private property, seize computers, documents and milking equipment, and fix CCTV cameras to trees. The authorities behaved as if milk were poisonous. Schmidt points out that in over 20 years, nobody has been harmed by his milk and that drug-dealing is treated more leniently than selling unpasteurized milk. The state has even threatened to take the farmers' children into care if their parents give them their milk. The judge in Ontario issued a permanent injunction against all those involved in the farm cooperative, prohibiting any further raw milk production without a licence, on pain of criminal penalties.

It is hard to understand the court's reason for closing the dairying operation when the judge only found Schmidt guilty of obstructing a police officer. It can only be by implication that the cow-sharing agreement was deemed unlawful, although I haven't read a transcript of the decision. The Canadian authorities are so determined to prevent the sale of raw milk that they will stretch the law almost to breaking point. The same thing happened in California in 2011, when the FDA raided Rawsome Foods, destroyed 800 gallons of milk and arrested its proprietor, James Stewart, for selling unpasteurized milk without a licence. Eventually the business was forced to close, while Stewart accepted a plea bargain and was fined rather than risk an expensive trial and a possible long American-style prison sentence.

In 2014, the UK Food Standards Agency came under some pressure to relax the rules relating to the sale of unpasteurized milk. It was argued that pasteurization had done much to degrade this once highly valued food. But the dairy industry, represented by Dairy UK, was vigorous in its opposition. It would prefer *all* milk to be pasteurized and was opposed to any relaxation of the rules. It argued that allowing the wider sale of raw milk to satisfy the 2 per cent of milk consumers who had shown an interest in it would cause 'significant food safety issues'. In other words, raw milk is dangerous and letting more people buy it and drink it would increase the risk of harm. People must be protected by the state from their own folly. This chimes with the view of generations of officials that pasteurization of milk is unarguable, and if all milk were pasteurized by law it would be a great advance in public health.

CHAPTER 9

My Little Herd of Heifers

WHEN I FIRST started farming, I knew next to nothing about keeping livestock. Apart from a few hens and some geese, I'd never had anything to do with the welfare of farm animals that belonged to me and for which I was solely responsible. My year at agricultural college was fine in theory, but it didn't tell me anything about the art of breeding domestic animals. I had no idea about the difference between good-quality cattle and ones that would cost you more to keep than they turned out to be worth. I had no idea that some would convert their rations into production while others would be so high-maintenance that they would eat their heads off for little reward.

Because I began farming as an outsider and in a small way, I was rather awed by big cattle, so I decided to start with some small ones, which I naively thought would be easier to manage. I had seen some Dexters in a field near Carlisle and thought they might just fit the bill. Little, black, tough and slightly hairy, these were Irish crofters' cattle, and I thought they would be hardy enough to thrive on my rather indifferent hilly grazings, where there was enough room for them to stay out all winter without housing.

I happened to mention to John, my farming cousin, to whom I turned for advice, that I was thinking of buying some Dexters.

'Dexters?' he scoffed. 'What the hell for?'

He turned to his son, who was a little older than me and who was passing across the yard with two full buckets of milk on his way to feed the dairy calves.

'Mick, d'you hear that? He's talking about buying Dexters.'

Mick snorted and carried on into the calf pens.

'They might suit my land and they're cheap to keep, and I think they're rather attractive little cattle,' I replied defensively.

'Oh, they might be pretty little things, but that's all they'll ever be. They're hobby cattle. They're cheap to keep because they're not worth spending money on. You generally get out what you put in.'

This was my first lesson in cattle breeding.

'If you want hardy cattle, you should try Galloways, or Luings. Or better still, Irish Angus crosses. Anything but bad-tempered daft little Dexters.'

It seemed too prosaically agricultural and unromantic to buy Aberdeen Angus heifers out of mothers that were commercial dairy cows, when I could have something from the Scottish Islands specially bred from the hardy Highland cow crossed with a Scotch Beef Shorthorn bull, beautiful and one of the oldest English breeds. And it would allow me the excuse of a trip to the island of Luing, where the Cadzow brothers had recently developed the breed. So I phoned them one day and was so encouraged by my conversation that I set off next morning to drive to Oban to catch the ferry. All on an impulse. I was twenty, ambitious and indestructible.

I don't remember much about the journey, or where I stayed on the way, although I do remember being excited to be off on such a trip, driving up the clanking metal ramp onto the little island ferry across the Cuan Sound from the Isle of Seil. I also remember the wild island scenery and the surprisingly comfortable stone house on what I thought of then as the edge of the known world.

The Cadzows proudly showed me their eponymous cattle on the island. The animals were hardy and hairy from their mothers and brown and deep cherry roan from their fathers, a sure indicator of succulent marbling fat running through the muscles and with a range of creamy colours to complement the red. Docile and thrifty, these were attractive cattle. Heifers were available to ship down to Cumberland. The Cadzows were enthusiastic about selling their new breed into the west Cumberland hills. I must have given the impression that I wanted to establish a pedigree herd and I remember they invited me to stay the night. I drove home the next day in sheeting rain, all the way from Oban to Lorton, only stopping for petrol, in my Renault 4 with the soft top that rolled up like a sardine tin. The little windscreen wipers barely kept pace with the deluge. I had energy to burn in those days.

I phoned my cousin next morning and told him I had decided to start a pedigree herd of Luings. I'd selected a few heifers and a bull – or rather, the Cadzows' herdsman had told me which were the best and I had neither the knowledge nor the confidence to argue with him.

'What? You've done what?' he shouted down the phone.

'*You* suggested I should consider Luings. So I've been to see them and I like them.'

'What do you know about pedigree cattle breeding?'

'Not much – as yet,' I replied, hurt by his straightforward reply. 'But I can learn ... and they're perfect for my rough land. They'll live outside all year and trample down the chest-high bracken.'

His next words were the second lesson I learned in my early farming life, which I have never forgotten.

'Stop trying to be a clever Dick. Are there any other herds of Luings in your part of Cumberland?'

I had to admit mine would be the first.

'Do you know why?' He didn't give me time to answer. 'Because they're not the right cattle for your land. Watch what people round about you are doing and do the same, only do it better. Ring up Jimmy Connon in Dublin and order thirty of his black bulling heifers. That's the best advice I can give you. And forget about exotic pedigree cattle. It's the quickest way to lose a good deal of money – money I know you haven't got.'

He gave me a phone number in Dublin and ended the call with 'His brother has three dress shops in Carlisle and Penrith and he's a dead straight decent fellow. Tell him I told you to phone him.'

The Irish voice on the other end of the line was courteous and delightful. 'If John's your cousin, that's good enough for me. How many do you want?'

I told him thirty and he said he was off on a buying trip 'out west' in a few days and would 'be honoured to act' for me – like an old-fashioned solicitor.

About a fortnight later – it was September time – I received a phone call.

'James Connon here. I have a consignment of fine cattle leaving Dublin on the tide tomorrow night. Your heifers will be part of it. The boat will dock at Silloth next day in the late afternoon. I've arranged haulage. The lorry is scheduled to deliver the beasts to you in the early evening, God willing, but the driver is instructed to phone you when he is due to leave the docks.'

John had warned me not to talk money with Connon because he might be insulted, so I didn't dare ask the price. I was told later that he worked entirely on trust. Nothing was ever written down. The amount he charged was based on the medieval notion of the just price. He paid a fair price for the cattle he bought from small farms all over Ireland and added a percentage to cover his costs and leave him with a reasonable profit. But he was concerned to ensure that his customers also made a profit. 'You should be content with your share of the profit and always leave the next man his share' was one of John's maxims that his Irish cattle dealer friend applied scrupulously.

The haulier phoned two nights later, just as it was getting dark. I knew him well. He was one of the brothers who owned the business. He had been carting livestock to and from farms all over Cumberland and Westmorland since he had been old enough to drive a wagon, and he knew every farmer for a hundred miles around. He even knew the layout of their farms and the field names. He had loaded two artics with black cattle from the boat at Silloth and estimated that he would be with me about ten o'clock on his way to Penrith. I told him which field I wanted them in and said I would keep an eye out for him so I could give him a hand.

It was after ten when the heavily laden lorry pulled up on the roadside beside the field gate with a hiss of brakes. The driver climbed down from the high cab and came round the side of the huge two-tiered lorry, cattle breath steaming from the vents in the sides, to let down the ramp.

'I put yours on last so they'd be first off. They're a nice quiet lot of heifers.'

We opened the folding gates inside and stood back in the road with our sticks and arms raised to make sure that when the animals ventured out they would turn into the field. One or two, braver than the rest, put their heads down and snuffled at the ramp with wet noses; once one had put a tentative hoof on the shiny metal, some of the others followed, startling themselves with their own clattering, jumping the last few feet onto the soft earth. Then the rest of the bewildered and bedraggled heifers came out of the dark recesses of the lorry and gingerly stepped onto English soil for the first time.

Once they had all been unloaded, they milled around, confined by the stone walls of the lane that led to the field that was to be their new home. They still had enough energy to trot round the boundaries in the moonlight and inspect the extent of their domain before resolving themselves into a herd on the hillock in the middle of the field. Their eyes flashed in the beam of my torch as I left them for the night.

Early next morning I slipped into yesterday's clothes and hurried up the field to inspect my new responsibilities. It was a still late-summer morning. Mist hung about the fell tops and the nip of autumn was in the air; a heavy dew had drenched the coarse tufts of foggage that had grown uneaten since I had taken a crop of hay a couple of months earlier.

A few of the heifers had found their way through the open gate into the bottom field and were grazing methodically, but the others had not dared to leave the familiarity of the higher field where I had left them the previous night. Some were still sitting down with their front legs neatly folded under them like dogs. Others were standing, conscientiously licking themselves clean of the muck and sweat and sawdust of their ordeal. They stopped what they were doing and stared at me as I approached quietly. I didn't want to alarm them. Considering what they had been through, they were surprisingly calm, self-possessed even.

Half a dozen sported big yellow flexible plastic ear tags with black numbers painted on them, and I presumed they had come from the same farm. Others were untagged, while one or two had a neat hole, about the size of a five pence piece, punched through one ear. This was in the days before it became compulsory to tag calves at birth in both ears.

Looking a bit closer, beyond these first obvious distinguishing features, I began to see them as individuals. This one had a wide muzzle and broad head – a sign of a good milker, John claimed; that one had a narrower muzzle with more of a tuft of black hair on the top of its head. Another was shorter in the leg than the others. Yet another was thicker set, with big soulful brown eyes. One thing they all had in common was that they were smaller and thinner than I had imagined – 'clapped like kippers', as John had warned me they would be. But they clearly formed a herd: all polled and black, some with little patches of white here and there – a blaze on the forehead, or a sock, or a few white hairs in the tip of the tail.

I was immensely pleased with my new purchases – they were the first cattle I had ever owned and the beginning of my herd. As I tiptoed back to the house for breakfast, relieved that, so far, they had all survived the journey, the sun was coming up on a new world.

I returned a couple of times that day to see how they were getting on, and each time they had spread out a bit further.

The grass was still flattened in the places where they had lain all night, but there were signs of more of them venturing into the bottom field during the day: here and there tufts of grass had been cropped, there were tracks in the grass left by their hooves, and sloppy cow muck.

The art of keeping domestic animals is to work with their natural inclinations, to guide their instincts and not to try to dominate or force them into unnatural courses; lead, don't drive, as John was at pains to advise me. Calm handling is the key to a relationship with any animal.

As the days passed, my herd settled down. I got to know them and they got to know me. Some would come when I walked amongst them and try to rasp me with their great slobbering black and pink tongues; others wanted to have their backs rubbed with my stick just above their tails. That's what a stockman's stick is, an extension of his arm. Others kept their distance but began to trust me, and no longer looked alarmed when I walked into the field. What was more gratifying was that they began to accept my collie dog when she came with me, so long as she stayed at my heel. If we got separated, some of the cattle would trot after her and try to chase her away. She resented this but was too clever to run off. She would lie still until they ventured close enough for her to bite them on the nose.

Gradually I came to realize that these were special cattle. John was right: they were just the ones for me. And if I hadn't had him to guide me, I don't suppose I would ever have known to buy them.

We had an unusually warm autumn that year and the grass grew well into October. After supper, I would put on my wellingtons, light my pipe and amble across the fields to see how the heifers were doing. They came alive in the early evening after a couple of hours resting to chew the cud, and started to graze with some determination. They seemed to enjoy cropping the wet grass after a day's rain. It must have tasted as good as it looked: clean and fresh, particularly when the late-evening sunlight reflected off the droplets of moisture making them sparkle like jewels. It was pleasure almost too painful to acknowledge just to stand in my field, leaning on my crook, collie dog at my feet, and watch them curling their tongues round the grass, tearing at it in great mouthfuls, as the last autumn light faded across my farm.

CHAPTER 10

To Hereford, to Hereford, to Buy a Big Bull

I F THEY WERE ever going to fulfil their purpose and breed beef, I had to get a bull for my little herd of heifers. They were sucklers, cattle kept for rearing calves for beef. These are distinct from dairy cattle, whose calves are a by-product of their main purpose, which is to produce milk. The dairy cow has to have a calf every year or so, to stimulate lactation, but it is reared separately from its mother. Almost immediately after birth, dairy calves are trained to take milk from a bucket. If they are left to suckle their mothers, it becomes the devil's own job to get them onto the bucket afterwards.

But my heifers were going to rear their own calves. They would grow much quicker suckling their mothers and would make much better beef. Besides which, being beef cattle, the mothers would only have enough milk for their own calf but none to spare.

'These are good heifers. What were you going to do about a bull?' John asked when he had pronounced himself satisfied with Jimmy Connon's consignment.

'I thought I'd use AI,' I replied a little tentatively. 'I want to use a decent bull on them and I can't afford fancy prices.'

Not for the last time, John scoffed at one of my suggestions.

'Catching these things bulling isn't going to be easy, but it'll be a piece of cake compared with getting them into the yard, on your own, to wait for the AI man. Forget it. You need a bull. A Hereford bull. They get small calves, which makes for easier calving.'

With typical generosity, he said he happened to be going to the bull sale in Hereford next week, and asked whether I might like to go with him. I never knew whether he was going because he wanted a bull, or whether he was doing it for me. He might just have wanted an excuse for a trip away. He loved travelling all over the country to sales and shows and NFU meetings. He went every year to Smithfield fatstock show, where he had exhibited cattle and sheep over the years; he knew just about every British farmer worth knowing from his days as NFU county chairman and he had an astonishingly wide circle of friends, national and international.

So it was that I found myself on the way to Hereford early one autumn morning, with John driving at his usual furious pace in his overheated car. I've noticed, over the years, that many people who work outside seem to like a hot car – and a hot house – and John was a notable example. He had booked us into a B&B. The only thing I remember now about our stay was being slightly perturbed by the sight in the morning of John's hairy, muscular white torso with his strong forearms and neck coloured deep brown with a farmer's tan, and his white pate leaning over the washbasin while he was shaving. He was over thirty years older than me and I hadn't been so close to an older man's naked torso before. My father was rather fastidious about nakedness, and he was blonde anyway, and much less hairy, so seeing John with his shirt off made a strong impression on me.

I remember it was raining when we left the guest house and drove to the mart. Inside the building, steam rose from the wet cattle and damp farmers and gave a pungency to the air that permeated the whole building. We made for the stalls to inspect the bulls tied up there, being pampered and brushed. The parts of a Hereford that ought to be white, principally the face, dewlap and socks, were being washed, blanched and preened. There was a buzz of excitement.

John was hailed loudly in Welsh by a florid, powerfully built farmer in late middle age, dressed in a fine tweed three-piece suit and highly polished brown boots. He said he had been a friend of John's for years and they were clearly delighted to see one another. He offered John an untipped Player's ciga-rette from a flat silver case, which John refused saying he'd given up because his body couldn't take it any more. But he would have a whisky with him if he were to offer. This was the first (at about nine o'clock) of many whiskies pressed on John and me by a cast of farmers and livestock dealers, most of whom greeted John as a long-lost friend.

Whiskies downed, we went back to the cattle, moving slowly behind the bulls, inspecting more closely those that took John's interest. Some he patted on the rump; he gently lifted the heads of others by crooking a finger through the ring in their nose, exposing their lower lip to inspect their teeth. He seemed to have a knack of calming them, because none put up much resistance. We picked our way amongst the muck and sawdust and people rushing around with buckets of hot soapy water, curry combs, brushes and towels.

One breeder in a long brown coverall coat approached John smiling. 'Well, well, John Scott! What the hell are you doing here?' He clasped his hand and wrapped his other arm round his shoulder.

John explained that we were looking for a decent, quiet bull, not too expensive, as this 'lad' was just starting out and he hadn't got much money.

'You've come to the right man. I have just the boy to do the job for you!'

He strode off expecting us to follow him, and stopped behind a bull that was sitting quietly chewing the cud with his eyes closed. He spoke gently – 'Come on, boy …' – and pushed at his rump with a boot. With some snorting and blowing the bull reluctantly got to his feet. There was nothing aggressive about this beast. He was as laid-back as a sultan.

'His father was like that. As quiet as a lamb. But he was very effective.'

John squatted behind the bull and weighing his impressive loose-hanging testicles, one in each hand, palpated them gently. As he straightened up he nodded to the owner to indicate his satisfaction and pushed between our bull and the one tied up next to him and, avoiding a frightening-looking downturned horn, grasped his ring and inspected his teeth.

'Have a look at these.' He motioned me to join him at the bull's head.

Gingerly I pushed through between the other side of our beast and the next one. Our bull's sleek muscled shoulders came up to my chest and he shifted his weight from one foot to the other as I pushed past him. John held up his head by his ring and pulled down his bottom lip to expose two big teeth in the middle of a row of six smaller ones, all of which evenly opposed the upper dental pad. He pronounced himself impressed.

While sound teeth are important in all grazing animals, they are not as crucial to cattle as they are to sheep. Sheep nibble off the herbage they eat, whereas cattle twist their tongues around the grass, pull it into their mouths and either tear or nip it off. A cow can feed satisfactorily without front teeth so long as the grass is long enough to wrap round its tongue.

Cows also feed relatively quickly, and with their sharp molars roughly chop the grass into pieces small enough to swallow into the rumen, the first stomach, which is a huge factory where billions of bacteria set to work to break down the cellulose into digestible substances. Once its rumen is full, a cow will find somewhere to lie with the rest of the herd and chew the cud. That involves regurgitating the partially broken-down contents of the rumen, and, using their chisel-like molars, grinding the material with a side-to-side motion, mixing it with up to 25 gallons of saliva and alternately swallowing and regurgitating again over a period of hours.

When the bacteria in the rumen have broken down the fibre into digestible amino acids, vitamins and minerals, the resulting substance passes into the next of the cow's stomachs, the reticulum. This acts in conjunction with the rumen, with the added function of trapping foreign objects, particularly stones, metal and bits of wire the animal might have swallowed, preventing them from moving further down the digestive tract and puncturing the wall of the intestine. Sometimes, if there is a greater than normal risk of ingesting ferrous objects, cattle can be made to swallow a specially shaped magnet, which will lodge in the reticulum and attract metal objects to it.

The food particles then pass to the third compartment in the digestive tract, the omasum. This filters out any large pieces and returns them to the rumen and reticulum for further processing; it acts as the gatekeeper to the 'true stomach', the abomasum, where hydrochloric acid digests the food and dead bacteria that have travelled down from the rumen, in a similar way to the human stomach. Beyond this are the two intestines, which treat the waste before it is expelled as cow muck, one of the most valuable fertilizers in the world.

We looked at some more bulls, all of which John said would do the job. All the while, more men hailed him and told him how delighted they were to see him. After that, we went into the auctioneer's office and introduced ourselves. And then the sale began.

The first bull was led into the ring on a spotless white rope halter by a man in a white coat. He encouraged it to walk with its nose in the air by pulling upwards on a thin rope attached to its ring. It kicked its front legs out when it walked, a trait that John found unappealing. Nonetheless, it made 800 guineas. John pronounced the next bull to be 'too ginger'. I didn't dare ask what was wrong with ginger, although I agreed that the colour of its coat did not attract me.

And so it went on, bull after bull paraded through the ring, until I became so bamboozled by Hereford bulls that I could barely distinguish one from another. I found some more attractive than others, but often when I expressed admiration for one, John would find some fault with it and I didn't dare bid. One or two we agreed on fetched a lot more money than I had to spend. Sometimes I would make a bid or two and then stop short of a winning bid because I was terrified of paying too much. The main problem was that I didn't know what I was looking for and so I didn't have the confidence to go all the way. Once or twice I pulled out of the bidding and missed a bull that John said afterwards I should have bought. But he hadn't said that while I was bidding, because he didn't want to influence me and get the blame for me buying a bull I later wouldn't like. The truth was that I would have liked him to buy me a bull, but I didn't want to ask him, and he credited me with being more decisive than I was. Also, I couldn't spot potential because I had no experience of watching an animal grow and develop. And as I could have taken just about any one of them home and been satisfied with it as soon as I had forgotten the others, I found it as hard to choose as I do from a restaurant menu when I can eat anything on it.

It was getting towards the end of the sale when John bought a bull for 970 guineas. He hadn't mentioned the animal before and I didn't remember inspecting him before the sale started. I rather liked him. He planted his hooves flat on the ground and stood well, and he looked up boldly at the crowd tiered around the ring. I still hadn't made a purchase when the bull whose testicles and teeth we had inspected came into the ring led by his white-coated owner. He plodded round and round, eyes bulging a little too much for my liking, but John said, 'Here's your bull' as the bidding started at the low figure of 200 guineas. It stuck there for what seemed like ages, until I held up the folded catalogue and nodded to the auctioneer, who immediately added 50 guineas to the price. I had only wanted to bid another 25, but I was too timid to correct him and I let it go. Later in my farming life I would have shouted, 'Twenty-five!' but I didn't dare draw attention to myself and I didn't say anything to John.

My bid seemed to break the impasse, because the bidding quickly went up without me, in 50-guinea jumps, to 650 guineas, where there was another hiatus, during which the auctioneer looked across at me expectantly.

'Give him another,' whispered John out of the side of his mouth.

Looking directly at the auctioneer, I barely had to move my head for him to get the message, and the bidding set off again, 50 guineas at a time. It stuck when I bid 950 guineas. I was feeling deeply uneasy about being carried along this high – and was wondering whether my bank manager would honour the cheque – when the auctioneer swept his arm around the tiers of benches, pointing with a little stick with a knob on the end at his other bidders and repeating, 'Nine hundred and fifty guineas ... I have nine hundred and fifty guineas bid for this exceptional bull. A bull that any self-respecting breeder would consider an honour to take home! Come on, gentlemen! Do I hear a thousand guineas for this fine bull?'

He glanced down at the seller, who was doggedly parading his unconcerned bull round the ring on a short halter, tweaking the cord through his nose to keep his head up and leaning in to his beast's neck as he plodded round. He betrayed no emotion. He was neither accepting nor rejecting my bid. The auctioneer was on his own.

He proceeded to recite his genealogy. 'He's out of a Princess cow by a Greengarth bull, gentlemen. There's fifty generations of breeding here and many of them I've sold myself through this very ring. Nine hundred and fifty guineas? Is that all he has to be? Nine hundred and fifty ... he's going, gentlemen ... I'm going to sell him ...'

I was sure I had bought my first ever bull. How the hell was I going to pay for him? I wished I hadn't bid so much. He *was* a nice bull, but was he *that* nice? I didn't really know.

'It's your last chance, gentlemen.'

He then pointed his stick at the under-bidder and said quietly, so as not to disturb the drama he had created and relished, 'I'll take twenty-five, if that will help ...'

I heard the man mutter, 'Go on then.'

'Thank you, sir!' declared the auctioneer triumphantly. 'Nine hundred and seventy-five guineas!'

Then he looked down at me.

My first emotion was relief that I was off the hook. I no longer had to shell out the best part of a thousand pounds I hadn't got. And I did not want to bid any more. Then I heard, through my confusion, John whispering, 'Make him a thousand! One more. He's a bloody good bull.'

'No. I don't want him that badly,' I whispered back.

I didn't say what I should have done – that I couldn't afford him. I just shook my head at the auctioneer and moved my outstretched hand, palm down, away from my body to indicate that I was done. But the auctioneer wouldn't accept it.

'Come on, sir. Round him up. A thousand guineas. You've stuck to him doggedly. You can't let him go now when another twenty-five will secure him. He's meant to be going home with you. He's *your* bull!'

The eyes of a hundred gnarled farmers were on me, willing the contest to continue. It was all right for them, it wasn't their money, and anyway they probably had a lot more of it than I did. But he *was* a good bull. And during the last few intense minutes I had even grown to feel he *was* my bull. I didn't want him to leave the ring belonging to the last bidder, a fellow I had begun to resent. And I did need a bull. In fact I needed one this week, and I wouldn't get another as good as this until next year's Hereford sale. And John was urging me to buy him ...

'Nine ... hundred ... and ... seventy-five ... guineas,' intoned the shrewd old auctioneer, scanning the assembled farmers for the slightest sign of another bid and sweeping his arm around the tiers of benches that raked up to the ceiling at the back of the round mart.

'Going once ... going twice ... going ...'

John's hand, grasping a folded catalogue, shot up and he said in a firm, steady voice, 'One thousand!'

'Thank you, sir! We have a fresh bidder. One thousand guineas.'

He knew that was the final price, because after the under-bidder shook his head grimly, he gave the mart one last quick scan and smacked the knobbly head of his stick onto the wooden lectern.

'One thousand guineas! He's going to Cumberland, gentlemen. One thousand guineas to John Scott from Cockermouth.'

The bull was led out and the hubbub resumed as the next bull came into the ring.

John got to his feet and nudged me with his elbow. 'Come on. Let's go and have a look at your bull.'

We went down the wooden steps and pushed through the throng of farmers leaning against the metalwork on the side of the ring, following the bull down the aisle between the pens. He certainly had an impressive backside; two muscled thighs flexed and rippled as he plodded along. His owner tied the rope of his white halter to the side of the pen and offered him a bucket of water, which he sniffed and disdained. The man then congratulated John on his purchase, saying he was the best Hereford bull he'd bred for some years and that John had got a bargain.

'Well, it's this young lad here who bought him. I was just acting as his agent.' And one of John's characteristic half-smiles lifted the right side of his face into a grin.

'Well, the luck will be for him then.' We shook hands on the deal and some folded banknotes passed surreptitiously from his hand to mine. I thanked him and put them into my pocket without counting them.

'He's a good bull. He'll give you some good calves. You can't beat breeding.'

I went off to the auction office to write a cheque I was far from sure the bank would honour and to arrange transport for both bulls for the journey from Hereford to Cockermouth. The haulier who had brought my Irish heifers had a wagon there and the driver was tannoyed. He accepted a commission to carry the two bulls home. He had a couple of others to deliver on the way and was anxious to set off. I left it to him to get them in his wagon and found John in the bar having a celebratory whisky with the seller of the bull he had bought.

My bull was called Jason and turned out to be everything the seller had claimed. He arrived on the wagon the next day and I turned him out with his new harem. He trotted down the field, where the heifers mobbed him, licking and nuzzling and jumping about like adolescent girls with a pop star. He proved not to be indifferent to their charms, because to my astonishment he mounted and served one almost immediately. I hadn't even noticed she was bulling. If I'd been relying on AI, I would have missed her and would have had to wait for three weeks before she came back into oestrus and could have been served again.

Jason was a gentle animal, easy to handle, with a lot of the steady character that the old ploughing oxen must have had. He accepted without demur being tied up in the byre for the winter. But he had one defect that was to seal his fate in the end, though it took some months for it to come out. In fact, it was not until late the following summer when he was out with the cows that it became apparent.

He was tremendously sensitive to the scent of a cow in season. He could detect it on the wind if it was in the right direction. You could see whenever he had the scent because it triggered the flehmen response, which is a sort of grimace that displays itself as an involuntary curling-back of the top lip to expose the teeth. Its purpose, in most cases, is to close the nasal passages so that the pheromones given off by a cow in season can enter the vomeronasal organ, or Jacobson's organ, an alternative sensory organ, wrapped in cartilage, that lies between the roof of the mouth and the palate. Bulls are only one of many animals to have this organ: it exists in animals as diverse as domestic cats, horses, hedgehogs, rhinoceroses and elephants. Its primary purpose is to detect two types of pheromones and hormones: those that act as sexual stimuli and those given off by animals in fear. The smell of blood triggers the discharge of the fear pheromone in cattle and is what causes them to panic at the slightest scent of it. The sexual pheromones mostly come from the urine of the female when she is receptive to the male for breeding purposes. A bull can detect impending oestrus two or three days before a cow ovulates, by tasting her urine.

The first thing my bull did in the morning when I let him out to graze was lift his nose into the wind, curl back his lip and take a long sniff of the air to see if there was anything that would need his services that day. According to our neighbour's wife, she had sometimes caught sight of her less-than-faithful husband doing the same thing as he left the house on a business trip.

Using his pheromone detector, a bull can smell a female on heat up to five miles away, and my bull was no exception. For a few months after he arrived, if nothing was doing in his own field of cows, he could be seen restlessly pacing up and down along the fence making low mooing noises. My neighbour's Friesian dairy heifers proved to be particularly alluring, and their proximity, even a few fields away, drove him to distraction. But he could do no harm because the couple of strands of barbed wire along the top of my fences kept him in. This changed one summer morning, when it all became too much for him.

'Hello?'

'Philip? That Hereford bull of yours is in with my heifers and he's bulled at least three! You'd better come and get him.' He put the phone down abruptly.

When I got there, he had paired up with a particularly flirtatious little heifer and they had separated themselves from the herd. They were standing together, his chin resting on her rump, and her flanks spittle-flecked; the hair at the top of her tail had been scuffed up and they were both agitated. It was quite clear what they had been doing and this little display was their post-coital rest.

There was no point in trying to separate him from these heifers while he was in this state – and anyway I knew I couldn't do it on my own – so I left him alone with his conquest and determined to go back later when it would be easier to prise him away from his paramour.

My neighbour phoned again after lunch.

'My heifers and that bull of yours are on their way to Lorton. They'll be in the village if we don't catch them.'

'Right! I'm leaving now.'

The little herd was grazing the wide roadside verge and nearing the village when we caught up, overtook and turned them for home. My neighbour drove in his old van and I walked alongside with my dog hanging back to head off any that might make a break for freedom. Out of the van window he said, 'He's bulled three. I didn't want Hereford-cross calves. These heifers are too good for that. I wanted to put a Friesian on them. They'll have to be injected and I think you should pay the vet's bill.'

'Er, hang on. He's a bloody good bull. I paid a lot for him. Calves by him would be worth something surely? It's not a complete dead loss. What if I pay half?'

He muttered something about my being a cheeky sod, but didn't answer me directly before I had to run back to stop one of his heifers breaking back.

When we got to his field, the heifers turned in through the open gate and we managed to separate my wayward bull from them and chase him off up the road on his own towards my farm, followed by my dog. But when my neighbour tried to close the gate on his heifers, it became clear how they had got out: the metal gate had been bent into a U shape and would not reach across the gap, and the chain holding it had been pulled out of its catch on the post.

'Your bloody bull has bent the gate.'

I couldn't deny it. Jason had pushed at the middle of it until it gave way under his immense weight.

'He doesn't know his own strength.'

But he did.

That was the beginning of his many romantic forays. Whichever field I put him in, he either deformed the gate if it was metal, or if it was wooden, he would push until it splintered in half. After two years, I had to sell him. I was sorry to see him go, because his calves were as good as his breeder had said. And he was as quiet as a lamb, despite his immense strength. But he had served his purpose. My heifers never had any trouble calving his calves.

The Hereford is part of the great tribe of red cattle that populated the southern counties of England in a sweeping crescent from Norfolk to Sussex, Devon to Herefordshire. It is thought that they all looked similar until a Dutch cross – introduced, it is believed, by the first Viscount Scudamore (1601–71), of Holme Lacy in Herefordshire – added bulk to the body and white finching on the belly, switch of the tail and rump, and in particular the face. Neither the coat colour nor the finching was fixed. The body colour ranged from red to yellow, white to grey or even silver, and the finching was 'fickle, freckled who knows how'. It took a hundred years of breeding for the 'bald' white face to become the distinctive badge of the breed's identity, which colour-marks all its progeny, and for the coat to become dark red to red-yellow.

The outstanding qualities of the Hereford were its thriftiness, docility, and good beefing qualities when they were sent into the Midland counties to be fattened after four or five years working the plough. The beefing qualities and strength of constitution ensured demand from the graziers of the world. The cows are long-lived, highly fertile, ready breeders and have a slightly shorter gestation period than other breeds. The bulls are renowned for their potency and determination to breed (a quality my neighbour did not seem to appreciate as much as I did). When coupled with the Hereford's outstanding capacity to lay down high-quality flesh on its ample frame, even when pastured on scanty fare, it is hardly surprising there are five million registered worldwide, from the arctic snows of Finland, to the dust and heat of South Africa, the hard grazing of South America and the droughts of the Australian Outback.

The remarkable thing about the breed is that its refinement into its modern form started in the 1740s, two decades before Bakewell began his work. Beginning with Benjamin Tomkins in 1742, a line of worthy West Midland and Welsh March families fixed the breed's valuable characteristics by selecting animals with the very qualities that by the 1780s caused it to spread out from its native county. It was the first English breed to be recognized as a true breed and it produced the finest English draught oxen ever bred. But as ever, consonant with the ancient truth of breeding, the Hereford's milking ability was necessarily sacrificed to the quality of its carcase.

William Marshall described feeding six-year-old Hereford oxen for beef, and could not help 'expressing some regret, on seeing animals, so singularly well adapted to the cultivation of the lands of these Kingdoms ... proscribed and cut off in the fullness of their strength and usefulness'. He remarked on the carcase of an ox, 'six years in the yoke', of which the flesh appeared coarse, but he 'had never eaten such high, fine, full-flavoured beef'. Arthur Young* approved of a 'lengthy period of work between maturity and the butcher's knife'. The great Northamptonshire and Leicestershire graziers went regularly to the Shropshire fairs to buy six-year-old bullocks and spayed heifers 'which had been drawing Salopian ploughs for three years'.

By 1846, when the first herd book was published, the perfected breed type as we know it today had been fixed. The clean white face and red coat colour had won out over the mottled faces and other coloured coats. It enjoyed great popularity with commercial graziers at home and ranchers in the New World. The home buyers went to the Hereford Michaelmas fair, which was 'not exceeded by any show of beasts in good condition in the Kingdom'. H. H. Dixon described the occasion of the great fair: 'On the third Tuesday and Wednesday of October the parade begins at Hereford station and extends right through the heart of the town' so that 'windows are barricaded against them and trapdoors burst in by them'.

Most of the home buyers were graziers from the northern and western Home Counties who sought stock to fatten for the Smithfield meat market. The exported animals stocked

---

* Arthur Young (1741–1820) was a prolific and influential eighteenth-century writer on agriculture. He wrote a series of valuable books describing his travels in England, France and Ireland. In 1784, he began the *Annals of Agriculture*, which continued for 45 volumes, attracting many prominent contributors, including George III, who wrote under the nom de plume 'Ralph Robinson'. Young was the first secretary to the government's Board of Agriculture, which involved him in the preparation of the comprehensive *General View of Agriculture* surveys of most of the counties of Britain. He was a hopeless farmer, but a superb chronicler of the state of rural life, farming and contemporary events at a crucial period in European history.

the prairies of North and South America and the beef ranges of Australia. The first Hereford went to America in 1817, but it was not until 60 years later that the breed began to be grazed in large numbers on the ranches of the Midwest. Its capacity to colour-mark all its progeny with a distinctive white face no matter what breed it was crossed with, ensured that it commanded a higher price than less readily recognisable breeds and crosses. It also nicked well with certain dairy breeds, the Ayrshire being probably the most successful, and also as a second cross with an Aberdeen Angus out of a dairy cow, as I happily discovered with my own herd.

The Hereford achieved early popularity because its improvers achieved the goals of all the eighteenth- and nineteenth-century breeders sooner than those of other breeds: to make the animal reach maturity early in its life and to redistribute the meat on the carcase from the front to the rear quarters. Maturity in a beef beast is described by Trow-Smith as 'the stage at which the animal begins to add fat to the foundation of bone, offal and lean meat or muscle which it has previously built up, in that order'. In unaltered breeds the final fat tended to accumulate in uneven patches. The early improvers strove to breed cattle in which the development of fat and lean meat – muscle – was a simultaneous process so that the animal reached maturity – became well and evenly fleshed – at an earlier age. In doing this, they also bred animals whose flesh became 'marbled'; that is, the meat was interlarded with fat rather than the fat being laid down separately from the muscle.

Redistribution of the flesh from front to back resulted in a square, blocky animal rather than a wedge-shaped one that had the balance of muscle at the front like a buffalo. The breeders aimed to change their cattle from animals adapted to the age-old dual purpose of drawing the plough and cart into ones that the urban market wanted, namely those whose carcase contained more of the valuable cuts of beef. The hindquarters of the improved cattle became deeper and well developed, the ribs well sprung and capacious, and the neck and legs shorter. In doing this, the back became shorter, but the process of reversing the wedge from back to front so that the hindquarters were the heavier part of the beast and re-lengthening the back had yet to be achieved.

There is another change that some of the breeders (perhaps unwittingly) made that has had to be reversed in the modern climate of grass-fed beef production. That is that some breeds became adapted to making use of the new feeding stuffs that were becoming available as the nineteenth century progressed. They were moving away from being pastoral animals, adapted to thriving on grazing, to being fed on concentrates, particularly oil cakes (the residue of certain manufactures, such as linseed cake) and brewers' grains. For example, the Aberdeen Angus (although it was not the only breed) became adapted to artificial stall-feeding (inside feeding on concentrated feed), in effect serving the valuable purpose at the time of transforming the cheap residues of crops brought across the seas from foreign lands into manure for the arable fields of Britain. The Hereford, on the other hand, never lost its predominantly pastoral characteristics, which ensured that it retained its popularity in the grazing countries of the world. When the emphasis swung back towards pastoral beef-rearing, Aberdeen Angus breeders found they had lost ground to the Hereford and had a good deal of work to do to re-emphasize the primary purpose of beef cattle: to make beef from grass.

Jason had done his job and got every one of my heifers in calf. But now that his wanderlust became irresistible and my heifers had grown into cows that would more easily give birth to bigger calves, I needed a bigger bull to breed more valuable beef cattle. I rather fancied a Devon, and that was where I looked next.

CHAPTER 11

Ruby Red Devons

THERE ARE SOME breeds that fit their landscape so perfectly that it's impossible to imagine any other type growing from the soil of that place. Some, writes Youatt, are 'beautiful in the highest degree'; there are others with an 'unrivalled aptitude to fatten' from grass and forage that can be grown on the farm, without expensive imported concentrated feed, and yet others in which fat and flesh combine throughout the muscles, marbling the meat and imparting the sweetest flavour. But there are few breeds that combine all these things. The cattle of north Devon, Ruby Reds, are just such a breed. They share the country with the South Devons. No other English county has two native breeds of cattle with such striking histories and each deserves a part of this chapter.

The red Devon has been common in the north of the county for so long that nobody can say there was ever a time without them. In Domesday, there is recorded such a remarkable concentration of *animalia* (this term does not include milk cows) in north-west Devon, centred around Barnstaple, that Trow-Smith is emboldened to speculate that it 'may point to the beginning of the slow expansion of the red Devon breed, which seven centuries later emerges into certain history as centred upon this area'. There are so few references in the historical records to colour, size or type of cattle that it is tempting to believe that the 'red' bull brought into Tavistock Abbey in 1366 as a heriot from a tenant at East Troswell was an early Devon.

The Pilgrim Fathers took Devon 'milch cows' as well as Dutch animals to the New World, where it was found that the English cattle withstood hardship better than the more demanding Dutch beasts and were 'much less trouble ... If any care be requisite, it is only for the purpose of giving them occasionally a little hay.' They didn't give as much milk as the higher-maintenance Dutch cows, but, wrote a Dutch observer at the time, they were much cheaper to keep and 'they fat and tallow well'. Nothing much has changed.

There is such a succession of references to a concentration of cattle in north-west Devon, in the breed's heartlands, that it would be reasonable to interpret them as showing that the red Devon has an unbroken ancestry as ancient as any British bovine stock. There is some evidence that in Elizabethan and later times they were smaller than they are today; for example, Trow-Smith mentions navy suppliers in Devon stipulating that the minimum weight for a fat ox should be 6 cwt. This he presumes (on no particular evidence) to have been live weight, which would be about half the weight of a modern mature red Devon beast. But it is hard to believe they could have been *half* the size they are today. The breed's heartlands are in the thousand square miles of upland and valley country between the Bristol Channel and Tiverton and between Barnstaple and Taunton. They were triple-purpose cattle. Although never givers of copious amounts, their milk was of an unrivalled richness (5 per cent butterfat for making butter and clotted cream) and they were reputed to stay in milk longer than most other cattle. Although their primary purpose was traction, at the end of their working lives (about the age of five or six) they were fattened for beef.

Until well into the nineteenth century in the West Country, long after they had been superseded by horses in other parts of England, oxen did most of the farm work, particularly ploughing. They were usually yoked two by two in teams of four, very occasionally accompanied by two horses. The reason for continuing with oxen was that the light, nimble horses of Devon and Cornwall were bred for negotiating hillsides with pack and pannier, rather than for the pulling power needed for the large wooden ploughs of the region. The same oxen always worked together in a pair, and each ox had a name to which it answered. Each pair had names that began with the same letter; the nearer one (left-hand side) was of a single syllable, while the far one's name had two or more syllables: such as Lark and Linnet or Belle and Beauty. The ox with the shorter name was closer to the ploughboy, and its shorter name was easier to say and hear, giving it a more immediate impact. Working collie dogs traditionally have short names for similar reasons.

Ploughmen and their boys commonly sang as they worked, to relieve the tedium of walking up and down the field all day, to put a spring in their step and also to communicate with the ox team, which responded to the human voice. West Country ploughmen's singing was unique. The ploughboy leading the team, walking alongside with his long hazel rod, with which he encouraged the oxen in their work, sang, or rather chanted, 'with unwearied lungs', almost from morning to night in a pure counter-tenor, just as if he was singing plainchant or recitative. And from time to time, as he directed the team, the ploughman would add in his part with lower notes in perfect harmony. Chanting encouraged and animated the team, just as armies sang as they marched, or sailors sang sea shanties. On still days, the ploughboys of Devon could be heard singing a long way off, and once heard, the haunting, ethereal sound was never forgotten.

Come all you sweet charmers and give me choice,
There's nothing to compare with a ploughboy's voice.
To hear the little ploughboy singing so sweet
Makes the hills and the valleys around us to meet.

Chorus:

For it's hark! the little ploughboy gets up in the morn.
Move along, jump along.
Here comes the ploughboy with Spark and Beauty,
   Berry,
Goodluck, Speedwell, Cherry,
And it's move along.
We are the lads that can keep along the plough,
We are the lads that can keep along the plough.

The old types of oxen, particularly the Devon, had stamina coupled with docility. They did well off what the farm would produce, and increased in value as they grew to maturity. It was said that a working ox gained a shilling a day, whereas a horse declined by about the same amount until at the end of its working life it was 'nothing but a hide and a bag of bones'. James Black of Morden in Surrey wrote in 1784 of his Devonshire ox team that the expense of an ox's keep was half as much as a horse and that he was worth two shillings a stone after his labour, whereas the horse at the end was worth five shillings for his skin.

Devon ox teams had something more: they were nimble and quick on their feet, and very rarely needed to be shod. At harvest time it was not unusual to see a team trotting at six miles an hour taking the empty wagon back to the field for another load of sheaves. No other cattle could do that. William Marshall in the late eighteenth century described them as the best workers he had seen anywhere, for although rather small, they made up for it with 'agility and stoutness of heart...'

A Mr Herbert, writing in the *Farmers Magazine* and quoted by Youatt in 1834, describes the Devon ox: 'Nimble and free, outwalking many horses, healthy and hardy, and fattening even in a straw yard, good-tempered, will stand many a dead pull, fat in half the time of a Sussex, earlier to the yoke than steers of any other breed, lighter than the Sussex; but not so well-horned, thin fleshed, light along the tops of his ribs, a sparkling cutter, and lean well intermixed with fat.' The cow is described as much smaller than the bull, which is still a feature of the breed, 'very quiet, the playmate of the children, a sure breeder, a good milker, a quick fattener, fair grass-fed beef in three months'.

The breed was largely left alone by the eighteenth-century improvers because there was nothing much that could be improved. It was a mercy for the breed that Bakewell had nothing to do with it. He is reported as saying that the Devon could not be 'improved by an alien cross'. If it had any defect, it was a tendency to be heavier and stronger in the forequarters than the more valuable hind parts as a result of its long association with the plough and farm work. But it was still an unrivalled grazing animal which found a ready market – too ready, because during the high prices of the Napoleonic Wars, many north Devon breeders were tempted to sell their best stock and breed from the worst. They adopted a ruinous short-sighted system that nearly did for their breed. 'If a calf which otherwise would be reared,' said Marshall, shows 'symptoms of a fattening quality, it is "bussed"; suffered to run with the cow, ten or twelve months … and is then butchered … Those which are of a nature to get fat at two years old, are *murdered*! Those which will not, are kept to breed from!'

Youatt thinks they were tempted to do this because until the end of the eighteenth century the north Devon farmers did not understand how superior their cattle really were. And even when they became aware of their good fortune, they were often slow to keep the best of the breed, which they 'retained almost in spite of themselves'. Thomas Coke of Holkham, the great Norfolk agricultural improver and self-publicist (later Earl of Leicester), early recognized the superior worth of the breed and by 1814 had 128 Devons on his own farms.

Despite its excellence, by the end of the eighteenth century, the breed's future hung in the balance in its own heartlands. Every Devonshire market day saw more excellent animals that should have been kept for breeding sold to the grazier or the butcher. Then, just in time, in about 1794, Francis Quartly of Great Champson near Molland on the southern edge of Exmoor stepped onto the stage. Fortunately there were still a few farmers who had refused to deplete their herds by selling their best stock, and wherever he found them, Quartly outbid the butcher and bought them to add to his herd.

In old age, Quartly confided to the Devon squire Thomas Dyke Acland, that 'perceiving that good animals were becoming scarce, [he] bought quietly all the good stock he could meet with and continued this over many years to improve his breed'.* Without a doubt, Quartly's skill and determination saved the Devon. The Quartly bull that stamped his character on the breed was Forester (1827), great-grandson of Quartly's Prize (1819), which had the oldest recorded pedigree in the Devon herd book. By the middle of the nineteenth century, nine tenths of the Devon herds were directly descended from old Quartly stock.

Francis Quartly was the youngest of three sons of James Quartly, who died in 1793 and left to him his leasehold farm and the herd that he had bred since 1776. The Quartly family had bred superior draught oxen at Champson since at least 1703. An inventory from 1725 of the estate of Henry Quartly of Molland details his herd: 8 cows and calves, value £28; 4 oxen, value £22; 4 steers, value £10; 8 two-year-olds, value £18; 6 yearlings, value £9.

Francis's brother, the Reverend William Quartly, was also a devoted breeder of Devons at his nearby farm at West Molland until in 1816 his stock and farm were taken over by his brother Henry. The valuation at the time put 11 cows at £127 (just over £10 each). The oxen were worth the most, at £13 13s. each. Henry continued to breed superior Devons

---

* T. D. Acland, 'On the Farming of Somerset', *Journal of the Royal Agricultural Society of England*, xi/i, 666ff.

until he died in 1840, when his two sons James and John succeeded to their father's and uncle's herds and maintained the standards. By this time the crisis had passed and other breeders arose, notably the Davy family (one of whom, Colonel Davy, founded the Devon Herd Book). By the 1850s, the Devon was second only to the Shorthorn in numbers in England, although there were nearly ten times the number of Shorthorns (4.4 million) than Devons (450,000).

Great Champson is still at the heart of the Devon breed. It is part of the Molland estate, owned by the Throckmorton family, a wonderful example of a landscape shaped by centuries of pastoral husbandry. The essence of the farming, the shape of the fields and the way they are divided has changed little for at least a thousand years. The happy combination of the breeders' skill and the effect of climate and terrain produces 'cattle beautiful in the highest degree'. But more than beauty, the cattle that grow out of this landscape are wonderfully capable of making the best use of the poorest roughage, calving easily, finishing to maturity entirely from pasture, and producing beef of the highest quality.

Three generations of the Dart family have farmed Great Champson since the Second World War, and the family has bred red Devon cattle in the area for much longer than that. Their herd is descended from Quartly stock, bred on the farm for over 250 years. The farmhouse is a fine example of a Devon yeoman's residence, with stone-flagged floors, dark Elizabethan panelling and, at the back, yet still part of the house, a huge dairy with stone sconces where clotted cream and butter were made and bacon and ham salted.

The great open fireplace in the kitchen, where whole logs were once burned, is now occupied by a wood-fired

Rayburn cooker. After showing me round the cattle in the outbuildings, William Dart and his son Richard invited me in for lunch of cold Devon beef and mashed potato. The slices of pink meat interlaced with fat were sweet, succulent and dissolved on the palate. This is beef as I remember it and as it ought to be. Grass-fed, natural and utterly delicious. If only more people could taste this, they would never eat bland grain- and soya-fed supermarket beef again. I ate slowly and reverently in homage to Devon and her ancient cattle.

A stream, diverted from its course in the deep valley behind the farm, and once powering a threshing mill, flows under one of the stone buildings and emerges at the bottom of the garden, where it is channelled along the contours of three fields on the side of the valley. Its flow, in open ditches, is controlled by sluices that allow the water to fill up the channel and then evenly overflow across the meadow below in a continuous sheet of water. Watering meadows goes back a long way. The art is to keep a trickle of water flowing through the roots of the spring growth, just enough to warm them, protect them from frost and give a first bite of sweet grass several weeks earlier in a cold spring. The water is kept flowing to prevent stagnation, never allowed to be still, but 'to enter the meadow at a trot and leave at a gallop'.

The huge acreages of water meadow that were once common in England have passed into legend, destroyed by the cost of labour, the loss of the old watermen's or downers' skills, handed down from father to son, cheap artificial nitrogen fertilizer and the difficulty of working the meadows with modern machinery.

Richard Dart bounced me round the fields in his Land Rover on a bright, blustery November afternoon. From the high land behind Great Champson, the panorama across north Devon spread out intensely green, with rolling fields topped by moorland and intersected by steep-sided valleys incised into the land by millennia of flowing streams. The fields and farms and villages were laid out in a glorious display of West Country beauty.

The Great Champson cattle had just been housed for the winter, more to protect the permanent pastures from their plunging hooves than the cattle from the winter weather. Red Devons are easily hardy enough to live outside all year round if they have some shelter from the worst of the storms that sweep across Exmoor off the Atlantic. Traditionally they had shelter behind the thousands of miles of hedges of ash, oak and hazel laid and woven into stock-proof barriers on top of chest-high turf and stone banks. On the higher land, beech is the only tree that can stand on the thin soils against the battering of the storms.

These Devon fields have been grazed by cattle and sheep since men first settled the land here. The farming has hardly changed, because grazing permanent pasture is the best and most economical way of using the land to get a living. And the Devon cattle have evolved through natural and domestic selection to make best use of this land. This is traditional pastoral agriculture at its finest.

At some point in the past – nobody really knows when – the Devon cattle in the south of the county took on a distinctly different form from those in the north. They became known as the South Hams cattle and are supposed to have arisen from a cross with the Guernsey or perhaps another Channel Island breed. There is also some evidence of an injection of blood from the humped Indian zebu cattle (*Bos indicus*) although this may have come via the Guernsey which has a similar genetic inheritance. In 1800, the painter and maker of livestock models George Garrard recorded in a footnote that a Mr Parsons said to him: 'I shall have the pleasure of shewing you my new Devons, which as a painter I know you will say have a finer claim to positive beauty than any you have yet seen – they are Calves got by an Indian bull given me by His Grace the Duke of Bedford, upon two year old new Devon heifers, and are as fat as quails at a month old ...'

There is also genetic evidence, because the South Devon is the only mainland British breed of cow that carries genes for both haemoglobin A and haemoglobin B. Every other breed only has a single gene for haemoglobin A. Both the Jersey and Guernsey carry both genes, and so does the zebu and other more southerly breeds. It is likely that the South Devon arises from a distant cross with one of these breeds – probably the Guernsey, because according to a report on the county's farming in the *Journal of the Royal Agricultural Society of England* in 1890, it was common practice to run a Guernsey cow with every ten or dozen South Devons to improve the milk.

In 1794, the Board of Agriculture was told that South Devon cows gave a decent yield of very rich milk and that both sexes grew to unusual sizes: the cows up to 16 cwt (800 kg) and the bulls up to one and a half tons (1,500 kg). The oxen were equally big. The breed was known for its extreme docility, which is still an attractive characteristic. Youatt rated their flesh as not as delicate as their northern cousins: 'They do for the consumption of the navy; but they will not suit the fastidious appetites of the inhabitants of Bath, and the metropolis.' He says they were reported as having 'more of the fourpenny and less of the ninepenny beef'.

The Shinner family, who farm near Buckfastleigh, have a herd that goes back even beyond the first herd book in 1891. Into living memory their cattle were kept as triple-purpose beasts: the bullocks worked and made beef; the females milked and worked and then became beef. Their milk was exceedingly rich – over 4.5 per cent butterfat – perfect for clotted cream and butter. It also attracted the same premium as the MMB paid for gold-top milk from the Channel Island breeds.

Devon clotted (or clouted) cream was made mostly in farmhouses in small batches, and with cheese and butter was a way of preserving milk. The method was to leave the milk to stand for 24 hours in 'a bell-metal vessel', says Youatt. Then it was heated very gradually until just starting to rise before simmering. From time to time the vessel was struck with the knuckle; as soon as it ceased to ring, or the first bubble was forming as it began to simmer, the milk was removed from the heat and left to stand for a further 24 hours, by which time the cream had risen and was thick enough to cut with a knife. The cream was then carefully skimmed off, and as much as was needed was saved, with the rest going to make butter and the whey to feed pigs. Youatt records that treating the milk in this way allowed five pounds of butter to be made from a given quantity, whereas ordinary churning only resulted in four; and the butter was 'more saleable, on account of the pleasant taste it has acquired ...' It took more goodness out of the milk, and left a thin whey, but 'it also gained a taste which renders it more grateful to the pigs ...'

Since South Devon milk lost its gold-top premium and more specialized higher-yielding dairy breeds have overtaken the dual-purpose types, and as the price for liquid milk is so low, the Shinners and other breeders of South Devons have concentrated on producing beef (or breeding stock) from their herds. In common with all the old British breeds, these are superb grazing animals. They calve in spring, to make best use of the flush of early grass, and the cows winter inside on silage and straw. And they are just as docile, if not more so, as their northern relatives. I was a little nervous of standing in a loose box with five huge un-ringed young bulls, and made sure I was between the nearest bull and the door. But both Robert Shinner and his son walked amongst them, petting them, tickling them with their sticks and showing off their good points, without the least concern. They really are quiet, and the Shinners have never even been roughed up by one of their bulls. Roughing up is when a bull pins you against a wall or other fixed object and rubs the hard top of his head up and down your body. He may not mean it aggressively, but a bull doesn't know his own strength and can do considerable damage if he becomes really enthusiastic.

The Shinners have just had to buy their farm from the Church Commissioners; it was either that or suffer the uncertainty of a new landlord, and they didn't feel they could lose the opportunity of a lifetime. I wondered whether the open fire in the kitchen, burning great logs in a huge fireplace, will survive the new owners. And it must be a deal of work for Mrs Shinner to keep the kitchen clean with a fire that is continually depositing fine ash everywhere. She said she rather regretted that there wasn't time to have 'the nice things in life', because everything was subordinated to the farm and the welfare of the cattle.

CHAPTER 12

Scotch Black Cattle

I T HAD NEVER occurred to me that my consignment of heifers was only the latest in a vast migration of cattle that has flowed through Cumberland from Ireland and Scotland across more centuries than it is safe to speculate. Although they have not come on their own feet for well over a hundred years, nothing else about their importation has changed much, except that there are perhaps fewer making the journey now than there have been for three centuries.

This migration of Scotch and Irish cattle to England may well go back into Neolithic times – more than 6,000 years ago. It is partly a result of the topography and climate of Britain, where the poorer soils of the north and west, which receive more rainfall (and have a consequently lower population), are more suited to stock farming than growing crops. In the south and east, the soils are more fertile, the climate more equable, and better crops and grass can be grown that will fatten grazing animals for the butcher. There are also more people in the places where the climate is better, and therefore the demand for food is greater.

Until the coming of the railways, cattle from the more remote parts of Britain had to travel either on their own feet, or by boat. And as they were one of the few commodities that *could* walk, cattle cost more to transport by sea than by land. Walking them was known as 'droving' – which has a romantic ring to modern ears – and was done by hardy, independent, reliable men, who always made their delivery, whatever the weather, unless prevented by *force majeure*. Nothing daunted them. Their word was their bond. Whenever they paid cash, it was on the nail – sometimes in gold. They were cunning and wily, knew every bend, bridge, ford, pothole and night stance along the hundreds of miles of their route, from the great cattle fairs of northern England and Scotland through the fattening pastures of the Midlands, all the way to the Great Wen (as William Cobbett called London), the maw that devoured everything that came near its centripetal force.

It has been estimated by one writer, Peter Roebuck, that over the centuries, Cumberland saw more beasts pass through on their way from Scotland and Ireland than any other county in England. Estimates of the numbers of cattle moving south seem to depend on which side of the Pennines the estimator knows best, because Kenneth Bonser in *The Drovers* gives the impression that the Yorkshire trade was the more important. But the Welsh trade was also large and valuable.

After I'd had my Irish heifers for a few weeks, and not having breached my cousin John's injunction against phoning Jimmy Connon to find out how much I owed him for them, I received a letter one morning in very elegant handwriting saying that James Connon presented his compliments and took the liberty of enclosing a note of his 'professional charges'. Attached to the letter was a wonderful old-fashioned bill in restful green curly printing, beginning with my name (spelled wrongly) followed by 'Dr. to James Connon, Grazier and Cattle Dealer' and his address. The 'goods' for which I was his debtor were 30 cross-bred heifers, at the price of £137 each, delivered from Ireland direct to my farm. Making a total of £4,110. I was astonished at the price. Even ordinary bulling heifers out of Cockermouth auction mart were fetching £250 apiece. I rang John to tell him I had finally got a bill and asked him if Jimmy could have made a mistake.

'Not him! There was a week when the price of stock fell through the floor. Don't you remember? He'll have bought them that week and he's passing on the good luck. Don't forget, you should always leave something for the next man to take his profit!'

I wrote out a cheque there and then and posted it with gratitude.

This is what it must have been like to do business with the old drovers in their heyday. Everything was done on trust. They took cattle from their owners and paid a smallish deposit in cash and the rest in bills of exchange or bonds, to be redeemed once the cattle were finally sold. Bankers were normally willing to discount the bills of exchange. That is, they would pay out against them in the knowledge that they would be honoured eventually. In the case of my dealings with Jimmy Connon, the whole transaction was done on trust. He didn't get me to issue a bill of exchange. He sent me cattle worth over £4,000, which he had bought (and presumably paid for, or promised to pay for) from numerous farmers in Ireland, entirely trusting that I would be good for the money when he sent me his bill. He based this on the confidence he had in my cousin, and at the time I had no idea how valuable this connection was or how honoured I was to be accepted into his web of trust.

The demand for meat from London animated the droving trade. And the ultimate destination, as it has been since at least the tenth century, of all those streams of beasts that plodded south and east was the great market of Smithfield – or Smoothfield as it was. Just outside the City, between the wall and the eastern bank of the Fleet river, Smithfield was a five-acre grassy field, once used for jousting and sports, close to water and grazing, where all manner of livestock was brought for sale. The names of the adjoining streets illustrate the breadth of the trade: Cow Cross Street, Cock Lane, Chick Lane, Duck Lane, Cow Lane, Pheasant Court, Goose Alley. The market has been in continuous operation for over a thousand years. And from earliest times, a considerable proportion of the fat cattle reared in the kingdom found their way to the metropolis via Smithfield. For example, in 1830, 159,907 cattle, 1,287,071 sheep and 254,672 pigs were sold through the market. These were for the sustenance not only of the people of the metropolis but also those of the towns and villages within an eight- or ten-mile radius – with some going to satisfy contracts to feed the navy. The volume of cattle sold had more than doubled since 1732, when it had been 76,210. The Monday and Friday cattle markets at Smithfield sold more cattle in one place than anywhere else in the world. By 1841, there was accommodation for 4,000 cattle (plus 25,000 sheep and lambs, 1,000 pigs and 300 calves).

Youatt estimated that the average consumption of meat in 1834 was 170 lb a year for each of the one and a half million residents of London – nearly half a pound each a day. He compared this with the 80 lb consumed by the average Parisian. 'But ours is a meat-eating population, and composed chiefly of Protestants; and when we remember that this includes the bones as well as the meat, half a pound per day is not too much to allow each person.' It's hard to understand why he thinks Protestants should eat more meat than Catholics.

Droving was a peculiarly British activity, partly due to the fact that the climate and topography made it hard to fatten cattle in the north and west of our island, which acted as a nursery for the breeding of cattle and sheep that would be sent south to be finished in lusher pastures. Also the kingdom was small enough for London to be the destination for the produce of the soil in the provinces, unlike Continental cities, each of which had their own hinterland. The English have always been great meat-eaters. Pyne's *Costume of Great Britain* (1806) has an engraving of a butcher (amongst many other trades) with the note below: 'It appears to be generally admitted that no people cultivate the art of breeding, fattening, slaughtering and preparing meat for the shambles, with so much care and success as the English; indeed the nature of English cooking demands the attention on the part of the butcher etc. as nothing can be more plain and unsophisticated or less likely to cover the defects of indifferent meat.'

Individual farmers had little to do with the sale of their animals, which they consigned to 'salesmen', who bought and sold as agents. And country drovers, who had often brought their animals a long way, carefully preserving their condition, handed them over to market drovers employed to pack the cattle into as small a space as possible. These men often had no interest in the welfare of their charges, which they treated with considerable cruelty such that a regulation was imposed at the beginning of the nineteenth century making each market drover wear an armband with his identification number on it, rather like police constables.

As the numbers consigned for sale at Smithfield increased during the late eighteenth century, the site became too small to hold all the cattle that were brought there, especially as the surrounding land and part of the original field had been built on. The animals had to pass through increasingly crowded thoroughfares, causing market days to be not unlike the running of the bulls in Pamplona.

Youatt describes the cruelty inflicted on the unfortunate beasts, 'barbarities which it would not be thought could be practised in a Christian country, if they were not authenticated beyond all doubt'. The overcrowding caused the drovers to resort to the most terrible methods to get the cattle to stand packed tightly in the space available. There were not enough pens on the field or room to tie the animals to the rails, so, starting at about two o'clock in the morning, the drovers would divide the cattle into 'off-droves' – groups of about twenty. The constables employed to police the market only worked during the day, so the identifying armbands made no difference to what happened during the night. The drovers would surround the off-drove and start to hit them on the head with their heavy sticks. To avoid these blows the animals would try to keep their heads low to the ground. At the same time, if they attempted to retreat backwards, they would be struck very hard on the legs and hocks or have sharp pointed goads applied to force them forwards. And if any dared to lift its head it would instantly be hammered by a dozen blows to the head or nose, about the horns or bony parts where there was least flesh to damage and where it would cause the most pain. The result was that however 'refractory, obstinate, stupid or dangerous at first', every bullock would sooner or later be disciplined to stand quietly in a ring – 'their heads in the centre, their bodies diverging outward like the radii of a circle' so that they 'may conveniently be handled by the butchers'. By breakfast time there would be twenty or thirty rings of cattle standing on the field in 'perfect discipline'.

The cruelty was redoubled when a beast had been sold and had to be separated from the circle. Having been bludgeoned to stop it leaving the protection of the group, it now became necessary to persuade the terrified animal to back out of the circle so it could be driven off by its buyer. It would be hammered on the head with great force, and often a goad was shoved up its nostrils to force it to leave its little troupe. At every opportunity it would try to rejoin the rings of cattle it was driven past, as it had been taught to do by the beatings it had suffered over the last few hours, seeking refuge from the violence. A bullock could be beaten out of as many as ten droves before the butchers' men could get it off the field and into the street on its way to the shambles that lined the market. On the way, the half-blind animal would run into or over or through anything in its path. Whenever a little ring was broken up to extract one of its members, the other beasts would be desperate to get their heads back into the centre as quickly as possible and re-form the ring for their protection. The circles were constantly being broken and re-formed, either by removing a beast or two, or by passing carts and drays.

There were many appalled witnesses to this unnecessary barbarity who tried to have it stopped. Secure penning would have avoided much of the horrible cruelty that turned the already traumatized beasts wild with terror, making them a great danger to the public as they were being driven through the streets to their slaughter. It is testimony to the inherent docility of the breeds of British cattle that they could be treated in this way without rebelling. It would not have been possible to deal with some of the more flighty Continental breeds like this. Many of the Limousin cattle imported into Britain during the 1970s were wild enough when treated quietly, but if they had been goaded beyond endurance they would have become murderous, and nothing but a bullet would have stopped them.

Charles Dickens wrote about the mayhem of Smithfield in *Oliver Twist* (published in 1838) and supported a campaign to have the market moved to bigger premises. In 1848, the new market for livestock opened in Islington, and Smithfield became the dead meat market that in part it still is today.

A significant element of the trade at Smithfield was in 'black Scotch cattle' – Galloways – which are the second most important polled breed to come from north of the border and almost certainly descend from the same stock as the Aberdeen Angus. Until about 1840, when the droving trade in south-west Scotland began to diminish, the black cattle – actually black or brown (dun) or occasionally white – made up the greatest part of the income (and capital) of that impoverished region. In the reign of William the Lion (1165–1214), the penalty imposed by the justices of Galloway for breaking the king's peace was a fine of twelve score cows and three bulls. This reflected the shortage of cash in a region where there was a large population of cattle. There are numerous references to the considerable numbers of cattle reared in this part of Scotland and driven south to England via Carlisle.

The climate, especially near the coast, is maritime and temperate and good for much of the year for the growth of grass. Before dairying (and sheep) pushed them to the margins, breeding beef cattle was the primary activity of farmers in the south-west of Scotland.

The Galloway type was once predominantly horned, although from early times there was a significant proportion of polled animals, which increased as time went on because they appealed to drovers, who were spared the risk of injury from horns. The polling is thought to have come from cattle that had been known in the Borders since Roman times, and probably beyond. The gene for polling is dominant in *Bos taurus*. So if two cattle are homozygous for polling – i.e. they have the gene for polling from both parents – a herd of cattle bred from them will be entirely hornless. It is also possible to cross a polled bull with horned cattle and have every calf hornless, because polling is dominant and horns are recessive.

From the middle of the eighteenth century, the breed was already more beef than dairy, although it milked better than its modern manifestation, with some strains yielding respectable quantities of milk. It was justly renowned for its beef conformation and quick feeding when transported to good English pastures, particularly in East Anglia, where it was driven in its tens of thousands to Norfolk graziers. In winter the animals were fed on turnips and grain to supply the winter market in the capital. The demands of the fatteners for a quick-feeding, even-fleshing beefing bullock, unaffected by dairying considerations, had a strong influence on the breeding and development of the type. But also no progressive breeders tried to 'improve' it, so it retained, unimpaired, its 'native characteristics', the most valuable of which is supreme hardiness, and the most notable its remarkable prepotency, which stamps its black colour and polling upon almost every other breed with which it is crossed. It does not mature early – five years is the average age when it is ready for the butcher – but when it is ready, its flesh is of superb flavour, marbled with fat, juicy and delicate. It is capable of being finished on grass alone. In the days when tallow and the hide were as valuable as the carcase, the Galloway would be worth as much as £2 a head more to the butcher than any other breed.

'Black and all black' is the mantra of the Galloway breeders. The animals look darkest in October when they have lost all traces of their brown 'calf hair', which usually returns in the spring and grows out during a summer of good grazing. It is said that the cows' long outer coat should be 'as wavy as bears' – though not wavy from lightness, because the coat has to shed the worst of weather – with plenty of wool underneath for warmth. Its coat is second only to the buffalo in thickness, and for that reason it tends not to lay down an outer layer of fat for insulation. This characteristic is attractive to the butcher, because what was once prized as tallow is now routinely discarded at slaughter.

Not all Galloways are black; there are dun, red, white and belted (both black and dun) types, with not much difference between them in performance. The distinctive belted type has attracted quite a following, as has the White Galloway, which is similar to the White Park in having black 'points' – that is, muzzle, ears and feet. There is also a strain that has red points. These white cattle do not breed to type, and if bred together for too long, the points will tend to fade, or even disappear. To maintain the colouring, they have to be crossed back with a black or red Galloway from time to time.

The Celts thought white cattle with red ears came from the Otherworld. They appear in many Irish heroic tales as fairy cattle, associated with the supernatural. In the Conversation of the Morrigan with Cu Chulainn in the Tain, 'the Morrigan came in the shape of a white hornless red-eared heifer with fifty white heifers about her and a chain of silvered bronze between each two of the heifers'. There is a recent interest in Riggit Galloways. These are a genetic oddity halfway between an earlier type with a white stripe along the back and tail and under the belly, and the solid colours of modern Galloways. Their unique finching is similar to the Gloucester and Longhorn in Britain, and the Austrian Pinzgauer. The historian of the Belted Galloway says that the finching is thought to have come from an infusion of Dutch Lakenvelder blood, 'probably in the seventeenth or eighteenth century'. But there is no evidence cited for this assertion, which seems to be based on the Dutch cow having a similar white belt. The indigenous, now extinct, Sheeted Somerset had an almost identical belt, and there are other old breeds that have similar marking. So it is just as likely that the belt is a genetic mutation inherent in the breed.

While its cousin the Aberdeen Angus was being developed into an early-maturing animal that responded to intensive stall-feeding, the Galloway remained a grazing beast for the hills and uplands. It stands bad weather and indifferent grazing better than almost any other breed and actually dislikes being housed in winter, being happier outside. Perhaps the main reason for its survival unadulterated through nearly three centuries is that a cross with a Cumberland White (or Whitebred) Shorthorn bull produces the Blue Grey, a wonderful example of the best effects of hybrid vigour. The heifers are highly prized as suckler cows, long-lived, thrifty, fertile and adaptable. And the bullocks make fine grass-fed beef.

Youatt declared that there was, 'perhaps, no breed of cattle which can be more truly said to be indigenous to the country, and incapable of improvement by any foreign cross than the Galloways'. The breeders were instinctive stockmen who understood what many farmers even to this day either ignore or do not recognize: that their breeding cattle should be rather *under* than *above* their pasture. In other words they should be capable of thriving on slightly worse land than they have. The Galloways had long been renowned for their capacity to tolerate fatigue better than most other cattle, a quality that was essential if they were to endure the long drive south to their finishing pastures. Their capacity, as the Reverend Smith succinctly put it, to 'grow thrive and fatten and to be reared at the least expence and afford meat of the most excellent quality' was what allowed them to bear comparison with any in the kingdom.[*]

The type was of a uniform quality that appealed to the drovers and the English graziers to whom they sold, particularly those from Norfolk, who prepared cattle for the London market. Inspection of one bullock in a troop was usually enough for the drover to judge, at a glance, the quality of them all. His skill was to know whether they would be 'good feeders' and 'sell best at the far end'. The chief Galloway sales were at St Faith's on 17 October and Hampton on 16 November. The drover would buy cattle at home, paying either in cash or more likely bills of exchange,

---

[*] The Reverend Samuel Smith's *General View of the Agriculture of Galloway*, written for the Board of Agriculture in 1810.

and send them off in droves of two or three hundred, in the care of a topsman, who was in charge of the gang of drovers – about one man to every 30 beasts – who accompanied them on their three-week journey to Norfolk. The topsman went ahead to secure grazing at night-stances, organize accommodation and make all arrangements necessary to ensure the trip was as trouble-free as possible.

Despite this, it was a hazardous, occasionally disastrous business. If disease struck, a drover could be ruined and be unable to honour his debts to the farmers. He did well if he could clear between 2s. 6d. (12½p) and 5s. (25p) a head for the journey. If he had the capital or the credit to handle a few large droves, he could make a good deal of money. If all went well, 1,000 head would leave him about £250. Walter Scott's grandfather was a highly respected drover in a substantial way of business, who made money from the 'Scotch' cattle trade.

Youatt quotes the Reverend Samuel Smith, who noted a 'peculiarity of character' of the 'greater proportion' of Galloway farmers that predisposed them to cattle dealing. It was either inherent or acquired from long experience of satisfying the English cattle trade, but they were in the habit of buying and selling for no reason other than 'the prospect of a good bargain'. When the markets were brisk, they would not keep a bullock more than a few weeks before succumbing to the temptation to sell it on at a profit. If the market was not favourable, they would hang on to it for a year or more until the right opportunity arose. Those few skilled in striking a bargain grew 'opulent' from their trading, but there were many others, 'tempted to embark in the trade, without either the talents or resources to carry it on', who did so nonetheless, 'frequently pursuing the road to ruin'. It seemed that the trade had 'all the fascination of the gaming table': the fluctuation of the markets, the sudden gains and losses, the risk, 'the idea of skill and dexterity requisite'. 'These excite the strong passions of the mind and attach the cattle dealer, like the gambler, to his profession ... He counts his gains but seldom calculates his losses.' Many farmers spent two or more days of the week at the auction, whether they had any business to do or not, and as a consequence, neglecting the attention that farming needed, incurred expenses greater than their income would satisfy. Their habits of dissipation sapped their desire to work and they became 'disqualified for any business or employment'.

In one of the Statistical Accounts of Scotland,* we find a description of the strange life of the 'lower kind of dealer' in cattle. 'He will travel from fair to fair for 30 miles around with no other food than the oaten cake which he carries with him, and what requires neither fire, table, knife, nor instrument to use. He will lay out the whole, or perhaps treble of all he is worth ... in the purchase of 30 or 100 head of cattle, with which, when collected, he sets out for England, a country with the roads, manners and inhabitants of which he is totally unacquainted. In this journey, he scarcely ever

* Surveys of life in Scotland during the eighteenth, nineteenth and twentieth centuries, containing information about the economic and social activities and the natural resources. The *First Statistical Account* (1791–99) was published by Sir John Sinclair, and *The New (or Second) Statistical Account of Scotland* was published under the auspices of the General Assembly of the Church of Scotland between 1834 and 1845. These two Statistical Accounts are among the finest European contemporary records of life during the Agricultural and Industrial Revolutions. A *Third Statistical Account of Scotland* was published between 1951 and 1992.

goes into a house, sleeps but little and then generally in the open air ... if he fail of disposing of his cattle at the fair of Carlisle, the usual place of sale, he is probably ruined, and has to begin the world, as he terms it, over again. If he succeeds, he returns home only to commence a new wandering and a new labour, and is ready in about a month perhaps to set out again for England.' There are others with 'wandering and unsettled habits' who 'job about from fair to fair without ever leaving the country'.

The Whitebred Shorthorn was developed in the wild Border country, the debatable lands of reiving days of the Middle and West Marches in north Cumberland, Dumfriesshire, Roxburghshire and Northumberland. They were bred from the Shorthorn and a type of white cattle that had existed in the Borders from early times, and have become a separate localized breed with a particular purpose. It soon became obvious to the instinctive Border stockmen that the cross with the Galloway produced an outstandingly valuable hybrid. In the late nineteenth century, blue-grey-coloured suckled heifer calves from hard Border farms, brought for sale at Newcastleton market, began to be recognized for their superior qualities as suckler cows on marginal land. They had the hardiness, longevity, carcase conformation and wide-ranging grazing habits of their Galloway dams, coupled with the milkiness, resilience and quality of flesh of their sires.

Their distinctive blue-grey colour is given to them by a genetic effect called incomplete dominance. When a bull that is homozygous (pure bred) for white hair is mated with a cow homozygous for black hair, the offspring will have a heterozygous mixture of black *and* white hair, rather than being either one or the other. As with all hybrids, mating the results of the first cross does not give reliable results in a second. For this reason, anyone who wants Blue Grey heifers has to repeat the crossing process for each generation. That is why people are prepared to pay about twice the price for a Blue Grey than they would for an ordinary cross-bred breeding heifer. Blue Greys are the bovine equivalent of Mules in the sheep world – the result of the genius of stockmen in the north of England, who wanted an animal that would cross with their resilient Galloways and fulfil the particular purpose of making the most of these extensive upland grazings. They scorned pedigree and fashion to get it, and created, above all, practical farmers' cattle, and none the worse for that.

The early Borders breeders credited with doing much to fix the breed's characteristics – Andrew Park of Bailey near Bewcastle in Cumberland, and David Hall of Lariston, Newcastleton, just over the border into Scotland – were both commercial farmers with the Borders stockman's eye for a beast that so characterizes the farmers of the country lying either side of the Roman wall. Although their land is bisected by the Scottish border, the people here have more in common, both culturally and tribally, with each other than with their compatriots to either the north or the south, with their history of 300 years of reiving and extortion. Blackmail originated here – the practice of demanding rent (*male*) as tribute in goods; black payment as opposed to white male, which was rent paid in silver coin. The Whitebred Shorthorn arose in the epicentre of this reiving country. Lariston was the stronghold of 'Noble Elliot of Lariston, Lion of Liddisdale' in James Hogg's stirring ballad of the reiving times, 'Lock the Door Lariston'.

When the Whitebred Shorthorn is crossed with a Highland cow, the resulting calf is a strawberry roan, which although not quite as good as a Blue Grey is a perfectly acceptable hardy upland grazer. Both the Galloway and its hybrid offspring will eat most herbage clothing the hills and uplands indiscriminately and are becoming appreciated (belatedly) as 'conservation grazers' – keeping invasive species in check and encouraging the growth of productive grasses, while treading down bracken, scrub and brambles. Many conservationists mistakenly held grazing animals to be the villains of the piece, eating out diversity and creating a monoculture, or overgrazing land and killing off unusual species. They are now coming to see that they were wrong, and increasingly accept that managed grazing is the best way of maintaining biodiversity and also producing natural grass-fed meat from the uplands.

CHAPTER 13

The Irish Breeds

Is trua gan ciar dhubh agam
Is trua gan ciar dhubh agam
Is trua gan ciar dhubh agam
Is Maire óna hathair!
I wish I had a Kerry cow,
A Kerry cow, a Kerry cow,
I wish I had a Kerry cow,
And Mary from her father!

Traditional Irish rhyme[*]

THERE ARE TWO remarkable breeds long native to Ireland. They are originally from the same Celtic stock (with admissions of Channel Island and Devon stock along the way), but at some time in the past they separated and the Kerry became a dairy animal. It is one of the first European cattle breeds to be bred solely for milk production and is the ideal crofter's cow: small (no more than 38 inches at the shoulder), thrifty, resilient and, for its size, one of the heaviest producers of high-quality milk of any breed. It will manage on almost any type of pasture and, if necessary, live

---

[*] Even the translation from the Gaelic doesn't make it any more intelligible.

outside all winter, growing a thick black coat that will repel rain and insulate it from bad weather better than any other dairy beast. Its light frame and comparatively large hooves allow it to graze the kind of sodden land common in Ireland, without doing too much damage. Three Kerrys can be kept to two larger cattle.

In the last few decades, the breed has lost out to the bigger commercial dairy cows, first Shorthorns and then Friesians and Holsteins, and is now a rare breed, kept alive in Ireland and North America by a few enthusiasts. The Irish government supports it by paying a grant of €86 to owners of five breeding cows or more for each pure-bred calf registered in the Kerry Cattle Handbook, and by keeping a herd at Farmleigh, the state's official guest house, which it bought from the Guinness family in 1999. A herd grazes the Killarney National Park demesne grasslands, and Murphy's ice-cream makers in Dingle, County Kerry, started using Kerry milk in 2006 to support the indigenous breed.

The breed is completely black (with a red strain that is not popular). The occasional cow has a little white on her udder. They naturally bear lyre-shaped white horns with distinctive black tips, reminiscent of an Irish harp, although most are de-horned now for safer handling. Bulls weigh up to 1000 lb (450 kg) and cows 900 lb (400 kg). Unlike the Dexter, they are renowned for their placid temperament; unusually for a dairy breed, even the bulls are considered docile. They are described as 'agile and active'; in other words, they have minds of their own, can jump barriers and sometimes are hard to keep in the field. They are easy calvers, having wide-set pelvic bones, and are long-lived, producing calves and milk well into their teens. They regularly produce 1,000 gallons and more of high-quality milk in a lactation, with 4 per cent butterfat and about the same amount of protein – ideal for butter-making. The globules of fat are much smaller than in the milk of modern commercial breeds – in this they are similar to the Gloucester – and therefore the milk is easier to digest, making it a well-balanced food easily tolerated by 'infants and invalids'. The Kerry's milk as butter and cream made up a large part of the diet of Irish country people, who like country people everywhere seldom ate meat.

The Kerry has been supported by the Irish government from early times. From 1888 to 1902, premiums were paid to encourage good bulls to be made available for breeding. This was to counter the sale of the best bulls to England and the retention of inferior sires in remote country places in Ireland. There was even a Livestock Breeding Act passed in 1925, which designated a Kerry Cattle Area in which only Kerry bulls could be kept. The regulations were later relaxed, but show the concern of the government to support their native breed.

Kerrys were once the dominant breed in Ireland, but despite the measures of support, they have declined to fewer than 1,000 in their homeland, with small herds in the USA, Canada and on the UK mainland. Recognizing their rapid decline, in 1951 the Irish Minister of Agriculture in a speech to the Dáil said he was 'irritated' by 'certain sophisticated farmers' bringing in Jersey, Guernsey and Ayrshire cows with no respect for the native breed, which was overlooked *because* it was native. If it had been an 'Andalusian cow people would be paying 200 guineas for it and would bring it in on a passenger liner. If we could make the Kerry cow as remunerative a business as the people of the Channel Islands have made of Jerseys and Guernseys we would have secured for the kingdom of Kerry a not insignificant source of income.' But the 'deplorable tendency of Irish people to look down their noses at their neighbour's son or their neighbour's beast is a perennial problem. If it's ours it is no good ... but if it is a Jersey or Guernsey everybody kneels down in front of it and says: "Is it not wonderful? Is it not lovely?"' Why, he asked, did the Irish people 'possess that supreme contempt for their neighbour's son or beast, but that son and beast conquers the world when they get outside Ireland'?

The Dexter is the other little black Irish cow, but it is more dual purpose than its Kerry cousin. Originally a crofter's house cow, a pronounced wild strain runs through the breed that makes some of them hard to handle. They're instinctively at home with other domestic livestock and are, like dogs, easy to train to the voice. They can be playful and naughty and go from obediently coming in for milking to chasing off a dog or a rabbit at full speed. Every herd has a leader, and if the leader can be controlled the rest of the herd will follow suit.

My neighbour had a herd of Dexters that he kept largely for amusement, but he also sold the beef from surplus young stock. He had two cows that were so wild they were dangerous. It was not wise to go too far into the field without leaving an escape route. One late summer evening we went to inspect the herd, which was grazing in the lower part of the field. One of the cows spotted us and made a low mooing sound, which alerted the rest of the herd. They lifted their heads in unison and gradually started to move uphill towards us, quickening their pace as they came.

'Come on! Make for the gate!' my friend said.

The herd was gaining on us as we moved purposefully towards the field gate about a hundred yards away. We started running and clambered over the metal gate just before one of the cows got there and ran into it with her horns down, lifting her head but not managing to toss the gate off its hinges.

'It's you,' said my friend. 'She doesn't like strangers.'

If the animals hadn't been so dangerous, it would have been funny: a few diminutive black and red cows, about the size of Shetland ponies, chasing a couple of grown men out of a field. But had they caught us, there is no doubt that at least two of them would have attacked and tried to gore us. They certainly did not persuade me to change my mind about the wild little things. They weigh about 800 lb and stand three feet high at the shoulder. They will live on vegetation that most cattle would turn their noses up at and are perfect for tethering in an orchard or on a bank or piece of steep waste ground, to eat it down instead of having to mow it. They will eat young nettles, green fronds of bracken and even seaweed. Their milk is even higher in butterfat than the Kerry, at 4.5 to 5.5 per cent, with certain cows giving milk of over 6 per cent. Two gallons of Dexter milk will make a pound of butter, whereas it takes nearly three gallons from most other breeds.

The Dexter is really a miniature version of the Kerry, created, according to David Low writing in 1845, by a Mr Dexter, agent to Maude Lord Hawarden (pronounced Harden) 'by selecting from the best of the mountain cattle of the district'. The cattle rapidly became more popular in England than in Ireland, and in 1886, at the Royal Show in Norwich, a three-year-old Dexter cow was shown in the 'Any Other Breed' class, while an English Dexter and Dexter/Kerry herd book was published in 1892.

However you look at them, Dexters can now be no more than hobby cattle. Even the long-legged version (they come in short- and long-legged types) is a dwarf, and as endearing as they are, they are never going to be commercially viable. When W. R. Thrower was writing about Dexters after the last war, he envisaged their being kept as a kind of house cow to supplement the rationing-restricted diet of people with a decent-sized garden and maybe a paddock or two or an orchard. How times have changed in 70 years! He gives the example of a woman who was living in a small house with a large shed adjacent, but no land other than a moderate-sized garden. She kept two Dexters in milk by grazing them in the lanes and neighbours' orchards. Her annual sale of milk gave her a useful increase in income. Another example is a 'professional man', living with his wife and four children on the outskirts of a large village in an ordinary house with a fair-sized garden 'vigorously cultivated'. His aim was to become self-supporting and provide his family with 'food they were otherwise denied', by having two Dexter cows graze an acre-and-a-half vacant building plot next door. Thrower does not say who is going to milk the cow, but it's implied that it will be the wife's duty.

He bemoans the acres of unused land visible even from a train journey from London to Manchester, and the wasted railway embankments and abandoned branch lines where a cow or two could be grazed 'by anyone with enterprise'. The grass burnt every summer to keep the tracksides clear would make more than enough hay to feed the cattle during the winter. He goes even further and suggests that a few cows could be profitably tethered to graze on the wasted verges of main roads, and the grass being 'scythed at council expense' could be gathered up by people living nearby to use for winter feed for a Dexter house cow. Water could be carried to the cows in old milk churns on a trailer hitched to a car, and the cows could be milked where they were tethered on the roadside.

'Quite a number of people live in a house with a paddock extending to three or four acres originally intended for the carriage horse of palmier days.' These paddocks had become a liability, and what better way of keeping the grass and weeds down than to own a few Dexters. It was a crying shame to him that wasted land was not dedicated to grazing a cow. 'A cow is designed to turn grass into milk (or butter) and grass is the best and most easily grown crop in the British Isles.' And for at least six months of the year cows will feed themselves on it. I couldn't have put it better or more succinctly myself.

In short, a man with no land at all could keep a few cows on ground that was producing nothing of nutritional value. At almost no expense he could have a gallon of milk a day and the meat from a bullock or two. In this way the annual loss of 500–600 small farms destroyed to make way for housing schemes, factories and roads could be mitigated. Thrower deplored the fact that rationing was still continuing seven years after the war, yet 'the profligate expenditure of land continues. A grim day of reckoning must come.' It is always invidious to predict the future, as Thomas Malthus would have learned had he lived long enough, and Dr Thrower was no different. Even though he could not foresee the sea change that was about to sweep across the farming world and domestic life, he was surely right about the wasted land and the food that could be produced from it. Had it not been for chemical fertilizers and pesticides and importing food from abroad, we would have needed this land lost to food production and the plucky little Dexter might have come into her own.

Moiled means 'bare' or 'domed' in Gaelic – as in Moel Famau (pronounced 'Moil Vammer'), the round bare hill and highest point in Denbighshire – and refers to this old Irish polled breed. In Low's time (1845), it was hardly known in England, but was apparently ubiquitous in Ireland, being most abundant around the River Shannon, and particularly at home in the 'drumlin country' of south Ulster. By the start of the twentieth century, the breed was confined to the three northern counties of Tyrone, Armagh and Sligo. Efforts by certain public-spirited leaders in Northern Ireland, notably Captain Herbert Dixon (later Lord Glentoran) and Captain J. Gregg, resulted in the formation of a society to promote it as a dual-purpose breed, particularly on the small hill farms in Ulster.

In 1929, the Irish Moiled Cattle Society laid down a pre-ferred colour standard of red or roan with a white stripe down the back, white tail and white underparts, like a hairy Hereford or Longhorn. In 1949, G. Perceval-Maxwell started the Ballydugan herd and imported a polled bull, Hakku, from Finland. Captain Gregg firmly believed that the Vikings had 'stolen' Moiled cattle from Ireland and felt this new bull was an appropriate return of blood from the old type. Hakku did much to revivify the breed until the government passed new regulations that prevented bulls being registered for breed-ing unless their dams had a recorded milk yield. As most breeders of Moiled cattle did not record, the breed went into decline, so that by the 1970s there were only thirty pure cows and six bulls left with two breeders keeping the breed going. Recently, support from the Rare Breeds Survival Trust, finan-cial incentives from the government and recognition of the value of native breeds adapted to the soil of their homeland

and able to turn grass economically into beef have caused renewed interest in the Moiled. It has even been given the accolade of a class at the Royal Ulster Agricultural Society annual show at Balmoral Park in Belfast.

Moiled devotees claim they are the only indigenous Irish cattle left on the island. But Youatt, writing in 1834, did not mention a polled type, although he has seven pages on the Irish version of the old Craven or Lancashire Longhorn, with which he was considerably impressed. Nor is there any mention of polled Irish cattle by early English writers. Sir William Wilde, Oscar Wilde's father, in a lecture to the Royal Irish Academy in 1858, 'On the Modern and Ancient Races of Oxen in Ireland', classified the Irish cattle existing in 1835 into four native types: the Longhorn, the Kerry, the 'Old Irish Cow' and the Irish Moiled, 'the Maol or Moyle, the polled or hornless breed, similar to the Angus of the neighbouring Kingdom, called Myleen in Connaught, Mael in Munster and Mwool in Ulster'. They were medium-sized, dun, black or white, rarely mottled, not bad milkers, remarkably docile and much used for the draught or the plough. He distinguished them from the 'old crooked-horned Irish' – which seem to be the beasts depicted in the engraving of 'Irish Cattle' in Youatt's book (p.181). These look remarkably like the Moiled – brindled, with a white stripe along the back and under the belly, but have the distinctive downward-pointing horns of a Longhorn.

This confusion might be put down to an ignorance of Ireland by English writers, because there seems little doubt that there had been polled cattle in Ireland for a long time. Polling appears spontaneously from time to time as a mutation in otherwise horned types. So a 'polled breed' is nothing more than breeding together two animals that have a dominant gene for polling and keeping on doing it by rejecting any animal with horns, until every animal is homozygous for polling. Thus it is hardly surprising that remains have been found in Ireland of a polled beast dating from AD 640. It is not proof of a polled breed; rather it shows that there was at least one polled animal. Even that most confirmed of hornless breeds, the Aberdeen Angus, regularly threw up horned animals until they were bred out by the early improvers. Moiled cattle most resemble a polled Longhorn, and it is more than likely that it is in this stock that they have their ancient origin.

To be fair to Captain Gregg, there are folk tales, nothing more, that claim that in their traffic between Ireland and Scandinavia, the Vikings took Moiled Irish cattle home with them. And there is a remarkably similar polled breed in Finland called Finn cattle, divided into three types, Eastern, Western and Northern, though they are of relatively recent formation. And so far as the evidence shows, the Vikings who raided and settled Ireland did not come from Finland. But as with anything to do with the origins of domestic live-stock, nothing can be proved, which is why we are inevitably thrown back onto folk memory, speculation and fancy. The misty history of the Moiled's origin fades into irrelevance compared with the modern story of its comeback amongst beef breeders in Ulster, who seem to have rediscovered a native polled beef breed that might one day rival the Scottish stock from their Celtic cousins across the water.

Over the centuries, Ireland has reared more cattle per acre than any other country. Its trade across St George's Channel over the centuries brought enormous numbers of store cattle to the fattening pastures of England, supplementing the flow of Scottish beasts and the trickle of Welsh ones. By 1663, it was estimated that 'for three years past there had been, one year with another, about 61,000 head of cattle brought over from Ireland in a year'. Tariffs were imposed to check the growing importation, which it was feared was flooding the English market, but they did little to reduce the rearing of cattle that had been taken up with enthusiasm by the Commonwealth settlers after Cromwell's Act of Settlement of 1653. They shipped them over in their tens of thousands, to be bought by the graziers of East Anglia. In 1665, 57,545 cattle were shipped over on the hoof to English ports. They were popular and profitable because after a winter of semi-starvation in Ireland, where they were expected to survive on what they could pull, they filled out and fattened very quickly on the lush pastures of England.

Such was the volume of imports that certain landowners who believed they were suffering from the competition – whom Pepys called 'the Western gentlemen' – proposed a ban on the importation of Irish cattle. The Duke of Ormonde violently opposed it on the grounds that it would hamper Irish economic recovery after Cromwell's devastations. The graziers of Norfolk and Suffolk feared for their business of 'buying lean cattle and making them fat' for the population of London, who also opposed the ban because they feared it would raise the price of beef. To the surprise of its opponents, however, in 1667 Parliament passed 'An Additional Act against the Importation of Forreign Cattel', prohibiting the importation of live cattle.

As is often the case with measures intended to protect agriculture, the consumer suffered because the ban did indeed raise the price of meat. The graziers were forced to buy store cattle at much higher prices from the West Country, Wales and Scotland. And it stimulated smuggling into ports in the west, which was connived at by the authorities. Ireland responded by slaughtering at home and selling carcases, and many farmers switched to keeping large flocks of sheep. Some ran to 20,000, mainly owned by English settlers, who had imported English breeds whose wool was said to be the equal of any from Leicestershire. The embargo lasted until 1863, even though it was observed more in the breach. For example, Youatt records that there were 79,285 live cattle exported from Ireland in 1812 (many to provision the navy during the French war) despite the embargo.

That so many store rather than fat cattle were exported from Ireland during these centuries is partly attributable to the unusual method of summering cattle on rented grass that obtained in many parts of the country. This reduced the amount of grazing available to each animal, with the result that few had the wherewithal to reach maturity.

In a *Survey of Londonderry* in 1814, the Reverend A. Ross described the 'mode of letting'. The cost of grazing a full-grown cow at three years old from May to November was known as the summ. A summ was divided into three feet. A year-old calf was a foot, and a two-year-old, two feet. A horse was five feet. Six sheep, or four ewes and four lambs, or 24 geese added up to a summ. If a summ was worth 6s., then the cost of grazing a two-year-old would be 4s., and so on. A summ on high land or poorer pasture varied, with the quality of the grazing, from 6s. to 10s. On fertile parkland grazing it ranged from £2 to £2 10s.

This was similar to an arrangement I came across on a piece of ground called Brackenthwaite Hows. The grazing was divided into 16 stints. Each stint gave its owner the right to graze one cow or six sheep. I had eight stints, so in theory I could graze 48 sheep or eight cows. But it was not clear when the grazing could be done. I rented the other eight stints from the other two owners so that I controlled all the grazing, but had they insisted on exercising their grazing rights, there would not have been enough grass to keep nearly a hundred sheep all year round, let alone 16 cows. And because three people had the right to graze stock over the same ground, there was no point in trying to improve it by cultivation or fertilizer unless we all paid a proportion of the cost. The other two owners were not interested in spending money on it, so it remained unimproved, an object lesson in the reality of communal ownership of land, a hangover from the days of open fields, and the reason for enclosure.

The Reverend Ross also attributed the inability of many farmers to feed their cattle beyond the store stage to the pernicious effects of gavelkind, the notion prevalent in Ireland of bequeathing a father's land and goods equally between his surviving children. This practice, 'so just and reasonable in theory, but so ruinous and absurd in practice, is interwoven in such a manner in the very constitution of their minds, that it is next to impossible to eradicate it'. In spite of every argument to the contrary, smaller landholders divided their land between their children until division was no longer practicable. In the course of two or three generations, a family would be brought to ruin.

Ross tells of a farmer with 30 acres of arable land and two sons. He divided his farm between them, with the result that neither could easily support his family. One son had four sons, and during his lifetime he too divided his 15 acres equally between himself and them, giving each three acres. The sons imagined themselves 'established landholders' with the means to marry, and promptly did so, creating four of 'the poorest and most wretched families that can be well imagined'. They had neither the land to produce the common necessaries of life, nor the money to get into another trade or profession. Ross bemoans the landlords who encouraged this subdivision as a means of increasing their political influence.

But the people themselves must also take some responsibility for their own impoverishment. At first blush, gavelkind seems to be based on a loving desire to treat all the children fairly. But if the result was to condemn them all to grinding poverty, where is the love in that? It is suggested that the deeper reason had something to do with a profound connection with land and a desire to have a piece that they could call their own. It is not just owning it; it runs deeper than that. It has to do with belonging and an attachment to the soil that is almost spiritual. So it is hardly surprising that they were willing to countenance the practical consequences of minute division.

CHAPTER 14

From Scotland to the High Plains of Colorado

OR TENS OF thousands of years the permanent grasslands of North America supported millions of the 'finest grass-eating creatures on four legs'. They were more like swarms of insects than mammals. No other herbivore has ever existed in such numbers as the American buffalo – or bison, depending on where you come from. It would have been easier to count the leaves in a forest than the number of buffalo in North America up to about 1860. It was estimated that at their most populous there were 100 million of the creatures roaming in great multitudes covering scores of square miles.

In 1889, William T. Hornaday, in *The Extermination of the American Bison*, reported seeing a herd on the Arkansas river he estimated to be 50 miles long and 25 miles wide, which took five days to pass by. 'From the top of Pawnee Rock I could see from six to ten miles in almost every direction. This whole vast space was covered with buffalo, looking at a distance like one compact mass, the visual angle not permitting the ground to be seen. I have seen such a sight a great number of times, but never on so large a scale.'

Some herds were estimated to contain seven million animals. By a large margin, there were more herbivorous animals on the North American continent 600 years ago than there are now or at any time since. The soil on the American prairies under permanent grassland before the buffalo were exterminated was some of the deepest and richest on the planet. Buffalo grazing over hundreds of thousands of years made it rich in humus, kept the grasses young and vigorous and anchored the roots to the soil. In the early 1700s, travellers described the magnificent silver pasture in Nebraska, where the grass grew eight to twelve feet tall, so high that even a man on a horse could not see over it across the prairie.

The herds were so huge that they frequently stopped boats as they crossed rivers, and overwhelmed travellers on the plains. Towards the end of their dominion they even derailed locomotives and held up trains by sheltering from blizzards in the newly opened railroad cuttings. So numerous were they that some Indian tribes believed the buffalo issued from the earth in an inexhaustible supply and that they could never disappear.

The early Spanish explorer Vásquez de Coronado, travelling from Mexico to Kansas between 1540 and 1542 in search of the mythical Seven Cities of Gold, saw 'an immensity of grass' grazed by vast herds of 'cows ... that it is impossible to number them ... there was not a day that I lost sight of them'.

Their range was continental. They were the biggest herbivore at the apex of the largest ecosystem outside the boreal forest that lies across the north of America from Newfoundland to Alaska. Superbly adapted to the extremes of temperature, from 110 degrees in summer to minus 30 in winter, they shared the land with tens of millions of other grass-eating, fertility-building animals: antelope, deer, jackrabbits, prairie dogs, innumerable beaver and coyotes, wolves, brown bears, and birds in flocks so huge that the legendary ornithologist John James Audubon described them as blocking the sun for three days. Over millennia these animals had created some of the deepest, richest soils on the planet: their dung had fertilized the earth and their grazing had pruned the grasses so that their stems tillered out in a thick mat of roots and vegetation, covering the soil and protecting it from frost, drought and storm.

There were hundreds of species of grasses, which covered a quarter of North America. The tallest grew on the eastern edge of the plains, across into Kentucky, giving way further west and at higher altitudes to the short grass of south-eastern Colorado. The swards contained a diverse array of other plants as well as grasses and sedges, all of which gave the grazing animals a range of minerals and nutrients that they brought up from various depths in the soil through their roots.

The buffalo migrated with the weather, roaming over the greater part of the modern United States and north into western Canada, grazing territory that ran 3,000 miles from the Great Bear Lake in Canada, south into northern California and Mexico, along the Gulf coast to Florida, then up the Atlantic seaboard almost to New York and west across to the Great Lakes. The land was one vast buffalo range between the Rocky Mountains and the Appalachians, from Minnesota to Louisiana. There were no fences and few natural predators to impinge on the dominance of the species. The herds practised a form of 'mob stocking', grazing an area intensively for a couple of hours and then resting to chew the cud before moving on to a fresh area.

Despite buffalo being close relatives of domestic cattle, the Indians never tamed them. Why would they want to when they could kill one whenever they needed to? Later efforts by Europeans were no more successful. The animals were said to be 'wild and ungovernable', and despite their bulk, they could jump nearly six feet vertically and run at 35–40 mph when they had to. The bulls weighed up to a ton and stood between five and six feet at the shoulder.

Buffalo had been exterminated from the eastern states by the last decades of the eighteenth century, but until the beginning of the nineteenth, both the native Indians and the buffalo were largely untouched in the 'Great American Desert' west of the Mississippi. Most of the area covered by the Great Plains had been owned by France (and Spain before that) and was 'a desolate waste of uninhabited solitude', wrote Robert Marcy after exploring the headwaters of the Red River. Not a place where people could live by agriculture, reported Stephen Harriman Long, the influential American explorer, in 1820.

In one of the most significant and audacious land deals in history, the emerging United States had acquired possession of this vast tract of land from Napoleon. It stretched from New Orleans, up the west bank of the Mississippi into what is now Canada and down the continental divide (the watershed between the Pacific and Atlantic oceans) into New Mexico, Texas and back to New Orleans. The purchase price for 828,000 square miles was $15m, an average of three cents an acre – an astonishing bargain even at the time. It sounds even more of a snip when converted to today's money, about

$250m. Napoleon's treasury was empty and he needed to finance his prospective invasion of England and keep his imperial ambitions afloat. The sale vies with the invasion of Russia as the worst decision the emperor made.

The prairies had been occupied for thousands of years by tribes of Plains Indians: Cheyenne, Arapaho, Pawnee, Kiowa, Sioux and Ute – after whom Utah is named. These formed loose tribal associations and occupied themselves fighting one another. The Ute's word for their dominant neighbours the Comanche – *kimantsi* – meant 'enemy'; the Comanche were the Lords of the Plains, extraordinary horsemen, magnificent fighting men and superb hunters.

The tribes may have spent much of their time in conflict, but they had at least one thing in common: their dependence on the buffalo, for clothes, tools, saddles, ropes, shields, utensils, meat – which could be dried, smoked, or stewed – weapons and shelter. They consumed or otherwise used every part of the carcase, including the organs, testicles, nose gristle, nipples, blood, milk and marrow. They generally preferred cows over bulls and particularly prized the meat from the hump, the tongue and unborn foetuses. Tepees were made from about 20 buffalo skins, dried, stretched and stitched together. They were light enough at 250 lb to be portable, yet weatherproof. The tribes' whole existence relied on the buffalo, just as the Eskimo's does on the seal.

After the Civil War, the Indian tribes soon came into conflict with the settlers who were pouring into the empty plains from the east. The Indians roamed as free as the wind that blew across land where 'there was nothing to break the light of the sun', as Ten Bears, the Comanche chief, explained in 1867 at the signing of the Medicine Lodge Treaty with the president of the United States, which was broken almost before the ink was dry. Killing the buffalo for their hides had already begun on a huge scale. Santanta, the chief of the Kiowa, asked bitterly at the Medicine Lodge council, 'has the white man become a child that he should recklessly kill and not eat? When the red men slay game, they do so that they may live and not starve.'

By the Medicine Lodge Treaty the Indians were promised perpetual hunting rights to most of the drier grasslands on the High Plains, south of the Arkansas river, while the settlers were allotted the wetter plains in the east. Yet within a few years of signing, the treaty had been broken by hunters who invaded the land and killed the buffalo in their millions, stockpiling hides and horns to sell back east. Seven million pounds of buffalo tongues were shipped back east out of Dodge City in one two-year period in 1872–3. One government agent estimated that 25 million of the beasts were killed at this time. Great heaps of bleached bones lay stacked at railroad terminals waiting to be sold, at ten dollars a ton, for fertilizer.

Greed and wanton destruction, unrestrained by either the national government or the western states and territories, and killing cows in preference to bulls, accelerated by the much-improved breech-loading rifle, ensured the buffalo didn't stand a chance anywhere on the continent. After the railroads came through, the railway companies used to slow the trains down to the same pace as a migrating herd so the passengers could clamber onto the roof or open the windows and blast away with the rifles provided on the trains for defence against Indians. The railway companies wanted the herds culled or eradicated because of the damage done to trains by colliding with buffalo crossing the tracks. A herd of buffalo could delay a train for days.

When the railroad came through the buffalo's grazings in Colorado and Kansas in 1870, it split the herd in two. The southern herd was confined to the Texas Panhandle, where it was annihilated within four years. Some people could see what was happening and proposed protecting the remaining herds. Some of the army officers who had been involved in the slaughter early on tried to stop it. William Cody ('Buffalo Bill') spoke up for preserving the buffalo because he could see the species was struggling. But those in high places in the United States were determined to subdue the Plains Indians, and believed that eradicating the buffalo would bring them to heel – rather as the US sprayed 'agent orange' on forests and farmland in Vietnam to kill foliage and deprive the people of food and shelter. For this reason in 1874, President Ulysses S. Grant vetoed a federal bill to protect the rapidly dwindling herds, and in 1875, General Philip Sheridan, in an address to a joint session of Congress, supported the slaughter as a way of depriving the Indians of their livelihood. Its destruction was a national ecological tragedy that set the stage for the later devastation of the Dust Bowl.

As the buffalo was being eradicated by just about anyone who could hold a rifle, the US army was doing the same to the Plains Indians, though unlike the buffalo, they did not go down without a fight. The most formidable opposition came from the Lords of the Plains, the Comanche; yet impressive as they were as fighting men, they were no match for the white man's weapons and ruthlessness. They were finally broken by a Texan army under the unscrupulous

General Sheridan during the Red River War of 1874–5. Six army columns descended on an Indian camp at Palo Duro Canyon. When the tribe fled, the army slaughtered their 1,048 horses, depriving them of mobility for the remainder of the war.

The last of the buffalo were wiped out within five years of the destruction of the Comanche. Sheridan told the Texas legislature in 1875 that 'lasting peace' could only be achieved if 'the Anglos killed, skinned and sold until the buffaloes are exterminated. Then your prairie can be covered with speckled cattle and the festive cowboy … forerunner of an advanced civilization.' It had taken only ten years to clear out the Indians and their buffalo; all that was left behind was mountains of skulls and bones and millions of sun-dried lumps of buffalo dung, which, while they lasted, the incoming settlers gathered to heat their sod houses and dugouts. The Great American Desert was about to become the Great Plains, where millions of acres of empty grassland lay open for the taking.

At the same time as all this devastation was being wrought, on the other side of the world, in a little corner of Scotland, a few far-sighted stockmen were engaged in the creation of one of the most successful and valuable breeds of beef cattle the world has ever seen. Not that they knew it, but it was a breed that in due course was destined to take the place of the beleaguered buffalo on the grasslands of the New World.

The Aberdeen Angus is one of the two breeds of polled black cattle – the other being the Galloway – to come out of Scotland, and is a testament to the remarkable skill of nineteenth-century Scottish stock-breeders. The breed originated in a polled strain of black cattle that had been around in Scotland since at least Roman times – and almost certainly much longer – part of the great tribe of Celtic black cattle that for many centuries had been driven into England from the Celtic parts of Britain: Cornwall, Wales, Ireland and Scotland. The two black Scottish breeds are distinguished by their being *homyl* – naturally hornless. Dr Johnson explained in 1775, in his *Journey to the Western Isles*, that the word meant 'humble', in the same sense that we refer to a humble – or bumble – bee 'that wants a sting'. Polled cattle have no power to gore.

One of the earliest references to polled cattle in Scotland is to a 'black homyl' in the Laws of Kenneth MacAlpin* who reigned from 843–860: 'If your neighbour's kine fall a fighting with yours, and if any of them happen to be killed, if it be not known whose cow it was that did it, the homyl-cow (or the cow that wants horns) shall be blamed for it; and the owner of that cow shall be answerable for his neighbour's damage.' It is hard to understand why the cow without horns should be blamed for the fighting, unless it is that a polled beast would have to be unusually aggressive to fight with a horned one and cause damage and must therefore be to blame for starting it.

In 1523, John Comyn of Aberdeenshire received 'unum bovem nigrum hommyl'. Hornless stock is known from archaeological remains in Roman Border settlements. And it is probable that a hornless strain of the Galloway and Angus breeds was selected to create the modern polled breeds. Polling

* Known as Kenneth I (810–58), who, according to myth, was the first king of the Scots and introduced an early code of laws.

Black and Dun Belted Galloways doing what they are bred for, turning indifferent moorland herbage into milk and meat and cow muck, balm to the soil, without which our marginal grazing would be much the poorer.

A plucky little Kerry cow with the breed's characteristic lyre-shaped horns with black tips.

The North American buffalo, 'the finest grass-eating creature on four legs'. Over millennia its dung made the American prairies some of the most fertile soils on the planet and its grazing created the thick mat of vegetation that stabilised the soil and protected it from drought, tempest and frost. At their most numerous, there were estimated to be a hundred million buffalo ranging from the Atlantic seaboard to California, from the Great Bear Lake to the Gulf of Mexico.

A blasted cottonwood tree stands beside the remains of a nester's sod-walled cabin, poignant witness to broken dreams in 'an immensity of grass' on the High Plains of Colorado in June. It is almost beyond imagination what the hundreds of thousands of people must have suffered who trekked into this wilderness in the hope of making a new life for themselves.

A 'mama cow' from Kit Pharo's Red Angus herd. These cows live as naturally as possible, like their precursors the buffalo, from the grass that clothes the prairie. They calve when growth begins in spring and their calves are weaned as it declines in the autumn.

A Red Angus bull on Kit Pharo's ranch. With a thicker, hairier pelt and small horns he could easily pass for a buffalo. The Red Angus absorbs less sunlight than his black cousin and so endures extreme heat better.

One of the feedlot pens at Burlington Feeders Inc. in early June, when the bare earth floors are hard and dry but the stink is still there.

Taken from 25,000 feet, crop circles in Kansas created by centre-pivot irrigation drawn from the Ogallala aquifer. Each big circle fits into a square section, with the smaller ones half- and quarter-sections.

The morning Longhorn cattle drive from Fort Worth Stockyards. A combination of 'living history' and religious rite, the twice-daily ritual keeps a perpetual memory of the myth of the cowboy, so potent in the American psyche.

An '80 inches TTT' cow in the Wampler T-Bar-W herd. She has distinctive Longhorn marking, slightly down-sloping ears hinting at *Bos indicus* ancestry, and the hind quarters deficient in beef that caused ranchers to be so disdainful of the Longhorn breed.

Pippin Star, one of the Wampler's young heifers.

The Wampler's stock bull about to serve an impressively horned cow.

The Osborne bull, now a cultural symbol of Spain, looming over an evening hillside in Andalusia.

A young Miura bull in the *dehesa* in December. '*Muy peligroso!*'

A representative sample of the numerous bulls' heads mounted on the walls of the Restaurant Postiguillo in Seville. Each bull has his name, date of death, his breeder and the matador who fought and killed him inscribed on the brass plaque below his dewlap.

White Park cattle in Jonathan Crump's herd. Note the black 'points' on the feet, muzzle, ears, round the eyes and the tips of the horns. This breed is of ancient British origin and would once have been found in most parts of the British Isles.

Landseer's well-known painting of white Chillingham cattle in a romantic Highland setting that bears no resemblance to their home domain at Chillingham. The affecting tableau represents the ideal family and played strongly to Victorian sentimentality: the bull watches over his cow, while she nurtures their calf with tender maternal care. This painting did much to fix in the popular mind the erroneous claim that these were the noble remnant of wild cattle that were once the 'unlimited rangers of the great Caledonian and British forests'.

was encouraged by the Scottish drovers because hornless cattle did less damage to one another and to those looking after them, and the breeders responded to the demand. It is desirable in cattle that are housed or fed in confined spaces and it obviates the need for de-horning, which is a routine operation in calves of horned breeds.

Very few remains of naturally polled cattle have been unearthed in Britain. One skull with no trace of horn cores and a prominent forehead, very like the head of a modern Aberdeen Angus, was discovered at All Cannings Cross in Wiltshire in a late Bronze or early Iron Age village dating from about 500 BC. Polled skulls of both the prominent-headed Angus type and the flatter-topped head of the modern Galloway were found during excavations at Newstead (Trimontium) near Melrose in the Scottish Borders, which suggests that the Scottish polled breeds are of ancient origin.

Cattle from Aberdeenshire and Angus had for centuries made up an important element of the hordes of 'Scotch black cattle' that were driven down to the graziers of England to feed on English pastures. For centuries in Scotland they represented a form of currency, making up for the lack of coin in circulation, and a store – often the only store – of wealth. At first there was little to distinguish between the Angus type in the east of Scotland and the Galloway in the west. They were both 'black Scotch beasts', and in fact the hairier Galloway seems to have been the more favoured due to its supreme hardiness and the finely marbled and highly flavoured flesh its carcase produced at about five years old after a summer on English pastures.

As the demand for flesh grew from the enlarging towns and cities, the more forward-looking breeders saw an opportunity for their superior beef cattle, which could use grass economically to produce a high-quality carcase. Hugh Watson (1789–1865) of Keiller in Angus was one of the first to see the potential in the type of cattle native to his part of the county, originally called the Keiller after the farm where he had become tenant in 1808. He is considered the founder of the Angus breed. The bull that started his line was Old Jock, born in 1842, the son of Grey-Breasted Jock. Old Jock was the first animal registered and therefore No. 1 in the Scotch Herd Book when it was founded in 1862. Another of Watson's notable animals was a cow, Old Granny (they weren't too imaginative with names), which was born in 1824, lived to be 35 and produced 29 calves. The pedigrees of the greater part of Angus cattle can be traced back to these two animals. They were better milkers than the modern Angus, which will do little more than rear her own calf. Watson described in a letter to the Highland Society in 1831 how he used his Angus cows to suckle five calves during a lactation: the cow's own calf would be born in January or February and suckled together with a bought-in calf until they could both be weaned onto hay, potatoes and gruel before being put out to graze on 1 May. The dam was then given two more calves to rear for three months before they too were weaned. Her lactation ended with her being given a fifth calf to suckle for veal.

Later, William M'Combie, MP (1805–80), of Tillyfour in Aberdeenshire, is credited with fixing the characteristics of the breed by blending the Angus with the neighbouring Aberdeen type into a superior beef animal. In doing so, he inevitably sacrificed the milk yield of the Angus to the high-quality early fleshing of the Aberdeen. M'Combie was a typically enterprising Scotsman, from a family of graziers and stockmen, who by the age of 19 was dealing in cattle on a considerable scale. He used the Bakewell method of line breeding, or 'close breeding' as it is called in Scotland, to establish an outstanding type that met with great success in England and, interestingly, France.

The 'strong black loam on the granite', as H. H. Dixon describes the land where M'Combie's best pasturage was situated, contributed to the quality of his cattle. Dixon describes being introduced to M'Combie's prize-winning ox in 1863: 'a little man would not be able to see him without assistance'. Lacking a ladder, M'Combie suggested they climb onto his manger to get a view of 'the vast plateau of roast beef'. 'Have you ever looked over more pounds?' He was 'beef to the root of the lug'. These fatstock farmers travelled huge distances to shows all over Britain and into northern France with their best animals. This particular bull travelled some 2,000 miles, won first prize at Garioch, and £40 and a gold medal at Poissy. At Liverpool, Aberdeen and 'on the grand tour', he took £130 in prizes over 132 weeks of peregrination. And he made £80 when he was sold for slaughter.

M'Combie's father had bought the 1,200 acres of Tillyfour in 1800 from the profits of his dealing in 'lean cattle'. When he was a young man in the middle decades of the eighteenth century, he had journeyed far and wide in Scotland, up to Caithness, Sutherland, Skye and the Islands, and brought home large droves of Highland cattle. There were no regular markets in these wild places and so the dealers would 'cry a market' – publish that on a certain day, at a convenient place, they would buy cattle. Although they could make large profits, they risked heavy losses, especially in spring, when the cattle were 'skin and bone' and had not the strength to make the journey south. Many died in transit. One night, after swimming the Spey – for there were no bridges in those days – M'Combie's father lost 17 'old Caithness runts' when it came on a severe frost after the cattle emerged from the river and they froze to death. 'Their bones bleached in the sun on the braes of Auchindown, for more than thirty years.'

M'Combie senior carried on a very large trade at the Falkirk markets and had an extensive business in England. He had a salesman who went to all the great fairs, particularly in Leicestershire, and sold the multitudes of cattle he consigned with drovers from Aberdeenshire. In one year he sold 1,500 cattle at the October tryst at Falkirk. He made £3,500 in two years at Falkirk alone. Most of his money was made during the Napoleonic Wars, when the price of cattle (and everything else) was very high. When the peace came, many people made heavy losses. One well-known dealer, George Williamson, was passing with his large drove through Perth when news of the peace was being tolled by the church bells. 'Old Stately', as he was nicknamed, often said that 'this merry peal was a sorrowful peal to him, for it cost him £3,000'.

M'Combie made money at the end of the eighteenth and beginning of the nineteenth century, when the expense of droving large numbers of cattle was, as he described it, 'trifling'. Drovers' wages were 1s. 6d. a day. They received no 'watching money'; there were no toll bars, and the roadsides and the commons 'afforded the cattle their supply of food'.

The breed was further improved by Sir George Macpherson-Grant, who had returned in 1861 from a Scottish upper-class education in England, at Harrow and Oxford, to his family estate at Ballindalloch, on Speyside, where he settled down to almost 50 years of dedication to the Aberdeen Angus breed. Macpherson-Grant also became an MP – for Elginshire and Nairn. He did not start from scratch, because he had inherited the oldest herd of Aberdeen Angus cattle in Scotland, which he improved with the purchase, in 1860, of a cow, Erica, from the Earl of Southesk's Kinnaird herd; she became the founder of his famous Ballindalloch bloodline. Macpherson-Grant was recognized as one of the greatest promoters and exhibitors of the breed. He won prizes at all the major national and international shows, at a time when there was intense competition in livestock breeding between farmers and breeders of considerable ability and intelligence. He took first prize at the magnificent Paris Exhibition of 1878 – staged to express France's recovery from her humiliation in the Franco-Prussian war seven years earlier – at which every leading British breed was exhibited.

These early improvers, all from a small part of eastern Scotland, elevated the Aberdeen Angus above the already renowned black Scottish types, into one of the greatest, if not *the* greatest, breed of grazing beef cattle in the world, an animal with a remarkable capacity to turn the flora of whatever pasture it finds itself eating into flesh. Extraordinarily thrifty, of medium size and vigorous, Aberdeen Angus bulls imparted pre-eminent fleshing qualities to every breed they were crossed with and were exported all over the world until well into the twentieth century. They established herds of first-quality grazing cattle on all the grasslands across the globe, from the plains of America, the prairies of Canada and the pampas of South America to the outback of Australia and the steppes of Russia. Black Angus is now the most ubiquitous beef breed in the US, and in Australia Anguses make up one in four registered cattle and a third of bulls sold at breeding sales. The breed is found in South Africa, Brazil, Denmark, Norway, Sweden, Spain and Germany, and remains very popular in Britain, particularly as a cross with dairy cows.

One of its most valuable qualities is resistance to extreme weather. The cows are good mothers, undemanding, adaptable and easy to handle, and the young cattle mature extremely early, with a high carcase yield of nicely marbled meat. A cross of the breed always improves the quality of the carcase of the offspring. Angus calves are small and thus easily calved, but hardy and lively. Until the colour was fixed as either black or red, it was indifferently black, brindle, dark red or silver yellow.

We have all come across people who seem to be able to eat anything and never put on a pound. Angus cattle are the opposite of that, the bovine equivalent of those unlucky folk who only have to smell a piece of cake for it to find its way onto their hips. And this is decidedly not about calorie intake. Cattle breeders know that the stuff about calories is nonsense: some breeds of cattle, and even individual cows within the same breed, simply use the food they are given more efficiently. And some cows can extract more energy from different foods than others. It is the same with people. So next time someone complains that they have tried to lose weight but can't manage it, have some sympathy; don't secretly scoff and think to yourself that they are just greedy and eat too much. They might well eat a lot less than you and still gain weight because they are, like an Aberdeen Angus, better fitted for survival on short rations.

*

In 1862, in the US, the Homestead Act was passed to offer free land to poor emigrants from abroad and from the east of the US to encourage them to follow Thomas Jefferson's dream of creating a land of yeomen farmers living on their own acres. They were lured to make the arduous journey by covered wagon (sometimes railroad) to stake a claim. In the west, there were 600,000 such claims across 80 million acres, much of which had been acquired under the Louisiana Purchase.

Any adult could apply for a 'quarter section' (160 acres; a quarter of a square mile) and become a 'nester'. There were certain nominal obligations: he had to occupy the plot for five years, plough a proportion for crops, and build a cabin of at least ten feet by twelve to 'prove up' the claim. Alternatively, he could buy the land for a nominal price of between $1 and $1.50 an acre.

The poignant remains of the dreams of thousands of pioneering families are dotted across the High Plains, their decaying cabins desiccated by the summer winds and relentless sunshine and broken apart by the winter's frosts, their impact on the land hardly greater than that of the Indian tribes they supplanted. A couple of cottonwood trees standing proud of the horizon and a grassy mound beside a damp place where there once was a creek are all that's left of many homesteads.

It is desolate enough driving out here across the miles of prairie in a modern four-wheel-drive pickup with air conditioning, but it must have sapped the human spirit almost beyond endurance to arrive on your allotted patch of wilderness, having tramped for hundreds of miles beside the horse-drawn wagon carrying your young wife and children, knowing that tomorrow was the first day of the rest of your life and this was your home. Unless you started to build your cabin – often with mud bricks, because timber was scarce and expensive – you would never progress from sleeping under the wagon or spending the winter in a tent. You had to feed yourself and your family from what you could grow or rear, and if you didn't, you would all literally starve to death.

Although much of the land had no surface water, the dry plains lay above the Ogallala Aquifer, a vast under-ground lake with an apparently inexhaustible supply of fresh water that had accumulated from melting ice after the last Ice Age. And fortuitously for the homesteaders, the self-regulating wind pump had been invented in 1854, providing a cheap method of raising the water. For about $75, a nester could buy a windmill kit; once he had bored into the aquifer, it would raise enough water to satisfy the farming needs of a 640-acre section without any power but the perpetual wind.

After the Indians had been cleared out, there was a cattle boom that lasted a decade or two. But it soon turned to bust. And by 1914, when the Great War began, most of the nesters that remained were struggling to survive. So it is hardly surprising that they enthusiastically took the federal govern-ment's advice to plough and grow wheat. This started an orgy of turning over sods that had never been turned before. The mat of grass roots was so thick that much of the plough-ing simply tore a slice off the top, ripping the turf apart. But a virgin acre would easily produce 15 bushels of wheat of 60 lb each – just less than half a ton – which would sell for about $2 a bushel. It cost about 35 cents a bushel to grow, so the profit on a half-section of 320 acres was nearly $8,000, a fortune in 1915, when factory wages were $25 a week.

By 1917, the national harvest of wheat, from about 45 million acres, was more than enough to feed the nation. Two years later, 75 million acres were planted and the plough-ing continued into the next decade on the 'greatest, gaudiest spree in history', as F. Scott Fitzgerald described it. Voices urging restraint were either drowned out or ignored.

If the farmers had had to rely on horses, the plains might have been saved, but the tractor and steel plough sealed their fate. With mules or horses it had taken 60 hours to plough, plant and harvest an acre. A hundred years later it took three. Once the railroad came, there was no check on the spree.

To complete the folly, the 1920s were unusually wet years. People who remembered the preceding droughts in the 1870s and 1890s were shouted down by the railroad men and the prophets of progress, who preached that rain followed the plough. Tearing up grassland caused atmospheric disturbance that would change the climate and make it wetter. And as if to prove them right, after most of the prairie sod had been torn up, lo! the rains did come – for a year or two.

In the five years between 1925 and 1930, five and a quarter million acres of native sward were busted up, and by the end of the decade a hundred million acres had been turned over. And the US government encouraged the whole thing all the way to destruction. The Federal Bureau of Soils proclaimed: 'the soil is the one indestructible, immutable asset that the nation possesses. It is the one resource that cannot be exhausted, that cannot be used up.'

By the summer of 1929, all across the plains, piles of unwanted wheat lay beside railway tracks as the price fell steadily. But farmers could not just reduce production; they had loans to repay on land and machinery, and the only way they could think to do it, with wheat prices half of what they were, was to produce twice as much corn. So in the autumn of that year, they redoubled their ploughing efforts, tearing up another 50,000 acres of land *every day* on the southern plains, land that had been under grass for millennia.

On 29 October 1929, the stock market crashed. It rebounded by the end of the day, but over the next three weeks lost 40 per cent of its value. The harvest that year exceeded previous years by some margin, and nobody wanted the wheat. It sat in heaps worth 40 cents a bushel, one eighth of its value ten years earlier. At that price it barely covered costs, so in order to try to keep going, once again the farmers applied the only remedy they knew: plant more wheat in more prairie than had ever been ploughed before.

Then in September 1930, a black dust storm blew up out of Kansas and rolled towards Oklahoma. People had never seen anything like it. It carried static electricity and felt like the swipe of coarse sandpaper on the skin. It was a harbinger, a straw in the wind, although nobody saw it; the beginning of the Dust Bowl, the biggest man-made environmental disaster in history, which destroyed a hundred million acres of some of the best land in North America.

The roots of the short grasses, which might have looked brown and dead in winter or in a summer drought, struck down deep into the sandy loam and knitted together, holding moisture even in a drought. But once the turf had been broken, the land became a desert. The soil offered little resistance to the wind, and the land that had briefly yielded the greatest wheat bounty in history was simply abandoned – as was the crop it produced. With nothing to anchor it, soil that for tens of thousands of years had nurtured a profusion of prairie life blew away.

The damage done in the Dust Bowl has never been repaired. It was impossible to replace the permanent grasslands that had stabilized the soil, fertile from 30,000 years of buffalo grazing. Nor was it possible to reverse the flight of the nesters from the land. There are fewer people living on the High Plains now than a hundred years ago, and not many more than when the Indian tribes ruled them. Mechanized industrial farming, consolidation of ranches and little alternative employment have caused an exodus to the towns that will never be reversed.

Meanwhile, those that have stayed have had to adapt to try to live with the devastation of the Dust Bowl, as we shall see in the next chapter.

CHAPTER 15

Feedlots and the Grazing Cow: the Maker of Fertility

A prophet hath no honour in his own country.

John 4:44

THERE ARE SOME things once seen or done that can never be erased or forgotten. Just as Adam could not un-eat the forbidden fruit, so seeing a feedlot, and above all, smelling it, is something that I fear can never be wiped from my memory.

The stench hit me as soon as I got out of the car, five miles out from Burlington, Colorado. It took a while to remember where I'd smelled something similar. But then it came back to me that it had been in the 1970s, at Wetheriggs, just across the M6 from Penrith in Cumbria, where the carcases of fallen animals were being rendered into pet food and glue. With the wind from the west and a good boil going, the smell permeated the whole town and beyond. Tighter regulations over emissions have now reduced it.

But not in Burlington, where tens of thousands of cattle – mainly the unwanted cheap bull calves of ultra-high-yielding Holsteins – are penned outside, with no bedding or shelter, in huge wooden-fenced stockades. Fed on a scientifically computed ration of corn (maize), chopped silage of triticale (a cross between wheat and rye) and soya bean meal, laced with growth promoters and antibiotics, they lie in their own muck (dry in summer and a quagmire in winter) without shade. Every mouthful is carried to them in feed wagons and once they are sent off to slaughter every particle of muck is either carted away or washed into big lagoons. This is a shocking place in its stark, uncompromising lack of sentiment. Everything is calculated to make the most return for the least outlay on the management and welfare of the animals penned there. By contrast, a fortune is spent on the latest machinery needed to grow and harvest the soya beans, corn and triticale that feeds the animals, and to chop it and cart it to them in the pens.

The often-heard justification for feedlots is that such intensive enterprises are necessary to feed the world's growing population. But all the vegetables consumed in the US are grown on just three million acres. And there are 35 million acres of lawn in America; they might make a start producing food from that before they claim it's necessary to confine cattle in feedlots. Americans are obsessed with mowing their 'backyards' – often within an inch of their lives. Even in the poorest places in the back country of West Virginia and Kentucky, where everyone seems to live in cabins and trailers, I saw dozens of fat people bouncing and wobbling on ride-on mowers over the grass surrounding their homes, with not a vegetable garden or poultry run in sight. Americans spend the smallest proportion of their income on food of nearly any nation in the world – 9.6 per cent – and they expect it to be like that, with the result that much of the food they eat is manufactured from denatured cheap ingredients and is clearly harming them.

The feedlot business treats cattle as nothing more than units of beef production. Holstein steers are not worth very much and will fatten to great weights on government-subsidized corn and soya, grown with genetically modified seed, using herbicide sprays that obviate the need for weeding. Large numbers of Holstein cattle can be fed effectively in feedlots, whereas they would never get fat from grazing alone – or at least not from the grazing available on the High Plains in Colorado. When oil and fertilizer are relatively cheap and the government subsidizes corn and soya, it is profitable to keep animals like this so long as water is easily and abundantly available for growing the crops to feed them.

On the High Plains this can be had by tapping into the Ogallala Aquifer that underlies this part of Colorado and stretches from Nebraska into northern Texas. Any disease caused by crowding cattle together in their own muck is kept at bay with antibiotics, and steroid growth hormones ensure that they make the most of the feed they consume and produce cheap beef to satisfy the American market.

Two thirds of all beef cattle and 90 per cent of cattle finished in feedlots in the US are given one or more of five hormones to increase their growth and feed conversion rate. Steroids have been approved in the US for cattle rearing since the 1950s. A voluntary planned phasing-out of certain antibiotics has been agreed between the FDA and agricultural chemical companies to try to stop the routine feeding of those drugs considered essential to protect human health. Not surprisingly, the feedlot producers have objected, saying they cannot manage without them. But when Denmark banned their routine administration to intensively reared pigs, though illness increased at first, as husbandry practices improved, the animals suffered no more illness than they had when they were given antibiotics routinely.

Burlington Feeders Inc. is one of 30,000 'feeding enterprises' in the USA, which range from 1,000 head of cattle to 100,000. Their critics nickname such places 'Cowschwitz'. On 1 January 2016, there were just over 13 million cattle in feedlots, with 80 per cent of those in feedlots with a greater capacity than 1,000, which are euphemistically called concentrated animal feeding operations (inevitably abbreviated to CAFOs). Seventy-five per cent of all beef consumed in the US comes from CAFOs.

When I visited, they were harvesting triticale (which makes up about half the diet of the cattle), with huge mowers racing across the circular quarter-section they were cutting. In semi-arid parts of the US, cropping land is made up of circles, watered by a centre-pivoted irrigation system. The Burlington enterprise crops nine of these circles – about 1,400 acres. I asked one resident why local people did not complain about the perpetual stink. He said that Americans generally do not complain about what their neighbours are doing, and the feedlot is big business and a sign of progress, both of which Americans tend to admire. The smell is just an unfortunate consequence that has to be put up with.

It is all highly mechanized and efficient. The cattle never go into the fields to graze, and are medicated with growth hormones, so they are docile to manage. Most feedlot operators take in cattle to finish for beef, and are happy to fill their pens with high-maintenance steers that eat a lot and take a long time to finish. They do not want cattle that fatten too quickly because they would have to find others to fill the vacancy. They also produce a uniform product, with no seasonal variations of appearance or taste, which is what the American supermarkets say the consumer wants.

In 2015, the US produced a record 14 billion bushels (392 million tons) of corn – the basis of both feedlot farming and the processed food industry, as well as ethanol production. There are about 90 million acres devoted to maize growing and 90 million acres to soya – half of which is exported, while a large part of the other half goes into animal feed. These crops are grown as industrial monocultures on a huge scale and are the foundations of cheap meat and processed food. They are underwritten by the US farm subsidy programme, which puts more than $25 billion a year into farming and encourages mono-cropping on large farms, particularly those growing these types of crops. Agricultural Risk Coverage and Price Loss Coverage are two important subsidies the American farmer receives to make up the difference between the market price and a guaranteed price set by the government. It is not hard to see why it is in the interests of the corn grower and the feedlot operator that the price of corn should be low.

All this is a far cry from what Kit Pharo is trying to do a few miles down the road towards Cheyenne Wells. When he was a young man, newly married and looking for work outside the rodeo season, he had the dubious honour of driving into the ground the first post to make the fence for the first pen on the feedlot. He is too diplomatic and lives too close to his neighbours to say anything critical about Burlington Feeders Inc., but the diametrically opposed way in which he manages his cattle and his pastures says it all.

Kit's ranch is halfway between Burlington and Cheyenne Wells. He 'raises seed stock' – breeds bulls – from his herd of 160 black and red Aberdeen Angus (or Angus, as they call them here) cows. His stock rearing is the antithesis of feedlot farming and sod-busting, with the simple aim of creating cattle that have characteristics as close as possible to those of the buffalo – hardy, self-reliant, frugal, resilient survivors that will thrive all year round on the available grazing and little else.

Kit's farming philosophy and reputation rest on the simple principle of matching his cows to their environment. This might seem like basic common sense, but to many who resist the truth of it, he is a lunatic. To others, he is a prophet crying in the wilderness that is modern American cattle farming. It comes as a revelation to those ranchers who keep cattle that are too high-maintenance for their circumstances. As Kit points out, the weaning weights of calves have steadily increased over the last 40-odd years, and yet most cattle farmers are struggling to make a profit. Those who were making a profit from calves weaned at 450 lb are now losing money with calves weaned at 600 lb.

Why? Because they are buying ever more expensive inputs to artificially adapt the environment to the cow, rather than the other way round. This has resulted in the average cost of raising a beef calf increasing from $384 in 2000 to $883 in 2014. And this is exactly the opposite of the way it should be. Growing and finishing cattle on grass results in a greater profit because there are no costs to set against it. And grass-grown beef is also better for you, because it contains far more vitamins, minerals and healthy fats – 50 per cent more iron, four times more omega-3, twice as much beta-carotene and much more vitamin A and E. Plus it contains no hormones, pesticides or antibiotics. As Kit points out, cattle are ruminants. They are designed to extract the energy from herbage through their four stomachs. They were not made to eat grains. And yet in feedlots their diet contains large amounts of grain. Grazing cattle is the only sustainable way to farm the grasslands of America – and the world – and produce food while building fertility.

On Kit Pharo's landing there is a romantic picture of a buffalo standing with his head down into a blizzard, caked with blown snow, defiant, proud, indomitable, the supreme survivor. The American buffalo is Kit's touchstone – the finest grazing animal that ever lived on the grasslands of the world – and the sales of his bulls depend on the extent to which his cattle emulate its qualities. With one crucial difference: the succulent meat from his range-bred Angus cattle is far superior to the buffalo's stringy dark flesh.

Kit sells bulls all over the US, as well as to Canada, Australia and Mexico, and franchises his genetics to other breeders so they can implant into their own cows embryos impregnated by his bulls. In all, he sells between 900 and 1,000 bulls a year. At his recent sale in autumn 2017, he sold 315 bulls – 184 Angus, 98 Red Angus and 33 of other breeds – at an average price of $4,900.

His cattle live outside all year round and are never fed, apart from hay in a 'doozy of a storm' when the snow is too deep for them to reach the grass below. His bulls do well wherever they go because there are few places where the conditions are tougher than where they are reared. Their lives are as natural as possible, with one cow and her calf having the grazing of about 30 acres of natural perennial grassland.

On the short-grass prairie of the High Plains, 150 years after the annihilation of the buffalo, the Angus is now the dominant grazing animal. By selecting calves from cows that can survive with little if any supplementary feeding, Kit Pharo has created cattle that 'fit their environment'. They can produce calves on 'what nature provides for them'. He describes these cows as 'adapted cows' that have 'very low-maintenance requirements'. In other words, they can maintain themselves because they have a greater capacity to extract the energy from their grazing than other breeds of higher-maintenance cattle.

The land here lies about 5,000 feet above sea level, and whilst not completely flat, it only rises a few feet to low ridges and bluffs and undulates into dried-up creeks and hollows, with little variation for hundreds of miles in every direction. There is an old plainsman's saying that you can sit on your porch in this part of Colorado and watch your dog run away from home for three days. In early June, the temperature is in the nineties all day, with a steady droughty wind blowing across the land, under a relentless sun.

Eastern Colorado had been suffering an extended drought for the last three years and there was little prospect of rain when I was there. Only the pale primrose-yellow flowers of the yucca plants seemed unaffected, standing sentinel on deep roots above their frieze of spiky drought-resistant leaves, which the Indians used to weave into baskets and utensils. The yucca spikes above the prairie like the prairie dogs that stand sentry on top of the hard-beaten mounds of earth above their burrows. They sometimes share these with the burrowing owl, which takes over the holes of other prairie creatures. When an owl finds itself threatened in its borrowed burrow, it imitates the sound of a rattlesnake, to fool and deter predators.

Apart from the heat and the relentless sunshine, this land reminded me of the high Pennines. I half expected a troop of Swaledale sheep to come out of one of the hollows. And Kit's cattle are managed in a similar way to the best flocks of Swaledales – only fed in the worst weather, and adapted to the terroir where they live as naturally as possible.

'How do you deal with calving problems out on the prairie?'

'We shoot 'em,' Kit replied, only half joking. They certainly select them so that they aren't breeding from cows that can't calve without outside help. They do rope calves that need treatment, but only *in extremis*; animals showing signs of weakness are culled, on the principle that if you breed from defective stock, your stock will become defective. Over 30 years or more, Kit has created a self-reliant type of cow that hardly ever gives trouble. He breeds for as much resilience as possible – even resistance to flies – to avoid veterinary treatment. His herd builds up immunity to disease, rather than him treating it when it arises. Keep the 'medium' right and disease will not take hold. He is with Béchamp and Bernard, rather than Pasteur, and believes in as little intervention as possible.

Drought, or *drouth* as they call it out here, is the hardest privation to deal with because it retards the growth of grass and hunger weakens the cows. It can be managed by reducing the size of the herd well in advance of trouble, or by weaning calves early to avoid them demanding from their mothers milk they could live without. Early weaning gives the cow the chance to put on flesh before the next punishing winter. It is better to sell half the herd than to weaken them all and risk losing the whole lot to starvation. And if the worst comes to the worst – and it often does on the plains – sell them all, put the money in the bank until the rain comes and then restock. At least that way you will have the wherewithal to buy more cows. Buying feed to keep a herd alive for a period whose end cannot be foreseen has left many a farmer with starving cows worth little or nothing and bankrupted him. If you wait until the last minute to sell, you will find that everybody else has done the same and nobody will want your thin cows.

There is truth in the land that cannot be ignored. These plains were meant for and made by grazing animals. Charles Goodnight, the legendary nineteenth-century cattleman, pronounced the Great Plains 'the richest sod on earth', and yet just over a century ago, the US government encouraged the destruction of the buffalo, drove off the Plains Indians and parcelled out the land to homesteaders, who were encouraged to turn the sod over and grow wheat. 'God didn't create this land around here to be plowed up. He created it for Indians and buffalo. Folks raped this land. Raped it bad.' As Kit Pharo says, 'Any time the government gets involved in farming, farmers become idiots.'

Ploughing up a million acres of permanent pasture on the Great Plains and killing the ancient grasses released vast quantities of carbon that had been stored for millennia in the plants and the soil. The annual ploughing that followed continued this release of carbon, until, having used up all the fertility, farmers have come to depend on chemical fertilizers to feed their crops. The deepest and most fertile soils on earth were, and are, under grass. These pastures are kept young and productive by grazing herbivores taking in the grass (biomass), digesting it, and excreting fertility in their dung and urine, which then grows more grass. But modern farming that depends on artificial fertilizers and annual crops removes this biomass, which is then lost to the cycle of fertility. Every civilization in history was built on deep, fertile soils created and sustained by grazing livestock. Any that neglected their soil did not last long. As for the claim that plants are more productive and ecosystems do not need animals, there are no healthy ecosystems on earth that are devoid of animals. Without animals, all soils grow sterile and eventually become desert.

You will hear it argued that cattle (and other herbivores) are inefficient users of herbage, in the sense that they excrete more of what they eat than they process. But that is precisely the point. An average cow eats 28 lb of plant-based food a day, drinks about 50 gallons of water and excretes 50 lb of muck. Her muck and urine are balm to the soil. They feed microorganisms, healthy bacteria, worms and a whole array of creatures whose actions create humus and increase the soil's water-holding capacity. A cow's manure is a magic product. Just ask any gardener what is the best fertilizer to transform soil into a productive medium. While compost is valuable, it is but a pale shadow of what has passed through a cow. The cow is an alchemist that transforms the base metal of sunlight, via vegetation, into the gold that is its manure.

A cow not only turns poor vegetation into rich manure and urine, but at the same time it prunes the old growth and stimulates the new, which over time encourages the grass to develop deep and extensive roots. Unlike most other plants, grass grows from the root, so pruning the tip makes its stems tiller out and increases its growth, which allows it to convert more sunlight into vegetation, which in turn decomposes, transpires and feeds the process of 'sequestration' of carbon. This is the much-mentioned, but little understood, 'carbon cycle', in which carbon combines with oxygen to make carbon dioxide. Plants take in carbon dioxide, absorb the carbon and excrete oxygen. And the whole process is hugely encouraged by grazing herbivores. Permanent grass is the marvellous natural covering of a third of the earth, and permanent grasslands are a more efficient store of carbon than any other ecosystem on the planet.

Most of the world's land, especially on the Great Plains, is unsuitable for ploughing: it is either too steep or the soil is too thin, or there is too little or too much rain. But grass in all its manifestations grows naturally over a large part of our planet. Grasslands are the lungs of the earth, kept in good health by their natural complement, grazing animals; a wonderful partnership that maintains the well-being of the planet. That is why grazing cattle must be at the heart of any sustainable farming system. And that is why Kit Pharo's Angus cows will, over time, repair the terrible damage done to this land and restore its health.

CHAPTER 16

The Texas Longhorn: an Ancient Breed in a New Land

The Texas Longhorn made more history than any other breed of cattle the civilized world has known. As an animal in the realm of natural history, he was the peer of buffalo or grizzly bear. As a social factor, his influence on men was extraordinary. An economic agent in determining the character and occupation of a territory continental in its vastness, he moved elementally with drouth, grass, blizzards out of the Arctic and the wind from the south. However supplanted or however disparaged by evolving standards and generations, he will remain the bedrock on which the history of the cow country of America is founded.

J. Frank Dobie, *The Longhorns* (1941)

HERE WERE NO cattle in the Americas before 1493. There were tens of millions of buffalo, but they are not of the same family.

On his second voyage to the Caribbean, Christopher Columbus brought some tough, thrifty beasts out of Andalusia to Santo Domingo, where the equable climate, vegetation and absence of competition favoured them. In a few years, these cattle, related to the race that now produces fighting bulls, established themselves on the islands. They grew taller, leaner and more powerful, with long legs that carried them tirelessly and very fast over rough ground and long distances, and long horns as a formidable defence against attack.

Modern genetic analysis has traced the Texas Longhorn's ancestry back to ancient times. It is a cross of 85 per cent *Bos taurus* (the progenitor of European cattle) and 15 per cent *Bos indicus* (Indian and African cattle). Its closest modern relatives are two of the thirteen Portuguese cattle breeds, the red-coated Alentejana, and the Mertolenga, which is a roan and broken-coated type that looks a lot like our own Shorthorn. They are directly descended from Christopher Columbus's cattle, which were from three different but related breeds: the Barrenda, Retinto and Grande Pieto.

As the Spanish conquistadors penetrated into Mexico and what became southern Texas in the early sixteenth century, they brought with them what they called their Caribbean Criollo cattle for fresh meat. Gregorio de Villalobos brought over from Santo Domingo 'a number of calves, so that there might be cattle, he being the first to bring them to New Spain'. Notable amongst these early adventurers and importers was Hernán Cortés, whom Keats has staring in wonder at the Pacific, 'silent upon a peak in Darien'. He is credited with introducing branding to the Americas to mark ownership and deter theft from the unfenced ranges. Cortés's own brand was a distinctive three crosses, which he used on his hacienda in Cuba and later on the great estate he developed in Mexico named Cuernavaca – Cow Horn.

These cattle came into a land with few enemies they were not equipped to deal with, and the vast open prairies provided all the sustenance they needed. By 1540, only two decades after their first introduction, they had so proliferated that when Coronado set out on his expedition from Mexico to look for the Seven Cities of Gold, he easily gathered up 500 cattle to supply food. On the way north, he abandoned any that were too weak to continue. Twenty-five years later, Francisco de Ibarra, travelling the same route, found thousands of cattle running wild. By the end of the century, a single owner in the province of Jalisco on the Pacific coast of Mexico was branding 30,000 calves a year. And in Chihuahua province, which borders Texas, there were single herds comprising tens of thousands of animals. At the inauguration of the viceroy in Mexico City in 1555, 70 or 80 bulls were rounded up for the bullfight from lands beyond all settlements, 'some of them twenty years old without ever having seen a man, *cimarrones*, outlaws, fierce and desperate for liberty'.

The Spanish custom in colonial times was to leave the males entire, and as a result, all animals were capable of breeding, which accelerated the growth of the herds. In any case, eating bull meat was supposed to be invigorating and promote longevity. Even as late as 1823, it was against the law in Mexico to kill calves for meat. Left to fend for themselves, augmented by further importations by Spanish colonists and missionaries, over time these animals formed huge semi-wild herds of wide-horned cattle that spread out over the more favourable parts of Mexico and west Texas.

By 1770, the mission of Espiritu Santo near Goliad claimed 40,000 head of branded and unbranded cattle between the

Guadalupe and San Antonio rivers. The unbranded animals were known as *mesteñas* ('mustangs'). After 1770, ranching up and down the San Antonio river declined rapidly under onslaughts from the native Indians – Comanches, Apaches and Lipanes – who slaughtered and stole cattle in 'unbelievable' quantities. Many cattle (and horses) were scattered across the countryside, where they turned feral and were sometimes hunted down by the Spanish for sport. The slowest were killed, while the fleetest and wildest escaped to breed. One Spanish rancher on the San Marcos river was so oppressed by Comanche raids that he was forced to abandon his ranch and those cattle he could not gather up. The ones left behind multiplied so rapidly that by 1833, Anglo-American settlers arriving from the north found the country 'stocked with wild cattle entirely free of all marks of ownership'. It was wrongly assumed that they were aboriginal feral cattle, rather than domestic cattle gone wild.

In the early to mid nineteenth century, when English-speaking colonists began to penetrate Texas, huge numbers of cattle were roaming in groups (not vast single herds) from the border with Louisiana in the east and the headwaters of the Brazos river in the west, as far north as the Red River and south to the Rio Grande. They were reported to be fearful of human beings, hiding in thickets during the day and grazing by night.

Then settlers from Europe brought their English Longhorns and Herefords, and others brought milk cows and oxen, some of which interbred with the Mexican and Spanish herds. In eastern Texas, the Cherokee Indians kept their own unique type of cattle, which they had to abandon when they were forced off their land. Even the Shorthorn was brought into the mix when Queen Victoria sent two cows and a bull from her own herd to Colonel Thomas Jefferson Shannon, governor of Texas, in 1848. They were landed at New Orleans and hauled in crates on ox wagons to north Texas so that they would clear the tick zone – where infective cattle ticks were endemic, and would have killed them if they had sucked their blood – before being allowed to touch the soil. But they did not meet with universal approval, being seen as too short in the leg. Some of the colonel's bulls were even shot dead on the open range by neighbours who wanted none of that 'squatty build' breeding with their herds.

But the characteristics of the basic cattle stock that became the Longhorn were so persistent that without fences to control their breeding, later imported types were simply absorbed into the breed without altering its basic character. These mixings resulted in the Texas Longhorn, a semi-feral, lean, flat-sided, multicoloured, self-reliant beast that lived off the land with no need for husbandry. They were not mature until about ten years old, at which age they carried an impressive spread of horn between three and four feet across. They were exceedingly long-lived, and resistant to severe hardship, parasites and disease. Although their lean meat was an improvement on the beef from the early Spanish Criollo cattle, it was still considered stringy, lean and too tough to be palatable, not unlike venison.

Gradually a Texas type of Longhorn diverged from the Mexican type and became widely accepted as superior to those that remained below the Rio Grande. This was partly as a result of the Mexican approach to selection for breeding, which was that if a calf was not doing too well and looked as if it would not make much of a steer for beef, it would be left for a bull. Whereas in Texas, the best bull calves were selected for breeding, and other males were castrated to make steers. There was also something in the terrain or climate north of the Rio Grande that grew rangier, bigger-horned and more powerful cattle than their Mexican relatives to the south.

By the time of the Mexican–American War in 1846, the Texas Longhorn had become a recognizable breed, distinct from its Andalusian ancestors, with a uniquely wide range of multicoloured coats, from black to blue-grey, yellow, brown, red and white, both solid colours and spotted, freckled, finched, brindled and even striped. It had also developed immunity to most of the endemic parasites and diseases that would kill many of the northern European breeds the ranchers were trying to introduce. The animals were lean, medium-sized (between 800 and 1,500 lb) and with a horn spread of up to five feet. But with the annexation of Texas into the US, the land became more comprehensively settled and fenced and the Longhorn lost its main value, which was the capacity to thrive on open ranges on indifferent vegetation. If they were pampered a bit more, the newer English breeds, such as the Hereford and Shorthorn, gained weight more quickly on improved pasture than the slow-to-mature Longhorn.

Wherever the Spanish had introduced cattle, numbers of them escaped to run wild in remote places. These *cimarrones*, outlaws, were there to be captured and tamed, or killed by settlers. Fresh escapees added to the breeding mix of the wild herds and replenished the stock of outlaws, which came to be known as mavericks – pronounced 'mav-rick' in Texas – after Samuel Maverick, a man whose name has passed into myth.

Numerous colourful stories arose to explain how these cattle got their name. One that blended fiction and fact has it that Maverick 'being a chicken-hearted old rooster wouldn't brand or earmark any of his cattle'. His neighbours all branded theirs – and as many of his as they could. Nevertheless, the old rooster claimed 'everything that wore slick ears' – had no ear-marks – or was unbranded. And when people spotted an unbranded animal, with 'clean ears' they would say, 'There goes one of Mr Maverick's beasts.' Then it got to be 'There goes a maverick.' By 1861, Samuel Maverick owned more cattle than any other man in Texas. After the Civil War, he claimed the hundreds of thousands of mavericks that roamed the Texas prairies – though of course he couldn't gather them up.

The least colourful story is the true one. Samuel Maverick was a lawyer in Texas who in 1845 reluctantly accepted a herd of 400 cattle in satisfaction of a creditor's debt of $1,200. The cattle were running free on Matagorda Island, supposedly in the care of a Negro family. By 1853, Maverick realized that the animals' caretakers were not looking after them, so he had as many as he could gathered up and moved to a range on the mainland. As most of his cattle were not branded, and most of his neighbours' were, they took the opportunity to 'maverick' them. This was not strictly stealing, because the 'custom of the range' allowed anyone to catch and mark any 'strays' with his own brand.

In 1856, Maverick sold his herd, still about 400 strong, to the romantically named Toutant Beauregard for $6 apiece, 'range delivery', which meant that the buyer took the stock uncounted, 'as they ran'. If there were more than he had paid for, he would be the gainer; if fewer, he would take the loss.

These were the only cattle Maverick ever owned, but as is the way of things, his name has become more indelibly associated with ranching than nearly all the cattlemen who gave their lives to it.

By the time of the Civil War, the Texas Longhorn proliferated across southern Texas into California. Its physical attributes and temperament inherited from its Spanish ancestor, had been modified by adaptation to the extremes of the American climate and terrain. Depending on the make-up of the herbage, a Longhorn cow could support herself and rear a calf each year on between 10 and 30 acres. And as there were millions of acres of open grazing on the Texas plains, and few predators that she could not deal with, a cow would normally survive to have between 12 and 20 calves in her lifetime. The native Indians tended to leave the cattle alone because they preferred to hunt buffalo, as they had always done, or other game; and the wolves stuck to their same atavistic prey, the less dangerous buffalo, rather than risk tackling fierce and well-armed Longhorns.

As the buffalo herds were being slaughtered to extinction on the prairies, the burgeoning herds of Longhorns spread out to take possession of the huge acreages of abandoned and ungrazed open ranges. By the end of the Civil War in 1865, there were estimated to be about 10 million Longhorns grazing across Texas, about a third of which were unclaimed. Nearly all able-bodied men had been involved in the war, either in the Confederate army, or against Indians, and there had been nobody to ride the far ranges and work the cattle that had never seen a rider or felt the rope. After four years of warfare, Confederate soldiers returned home destitute. Almost the only occupation available to them and the youths who had grown to manhood during the war was catching and branding maverick cattle, which belonged to whoever could get to them first and had enough fighting men to keep hold of them.

Young men would be hired as 'brush poppers'; for $10 a month plus board, they combed the sage brush, popping out cattle and rounding them up for the drive north. One rather facile saying arose that all it took to make a cowman was 'a rope, nerve to use it and a branding iron'. But there was more to it than that. It was one thing to brand mavericks, but an entirely different thing to keep the cattle you'd claimed and get them to markets a long way distant. And when the beasts were plentiful and relatively easy to rope, the price was low.

But if a man could brand and hold the cattle, there was money to be made. The Civil War had depleted stocks of European cattle in the east, and a steer that was only worth $4 in Texas would fetch up to $50 in Chicago, Cincinnati, Dodge City or one of the meat-packing towns in the north – that is, if the beast could be got there. That was why for rather more than two decades after the war, vast herds of Longhorns were gathered up and driven north by tough and desperate men along cattle trails. Cowboys, as they came to be called, were paid about $30 a month, plus keep, for the four to six months it took to drive a herd the 1,500 miles or so to the railheads in the north. A fair day's progress for a herd was about ten miles. They were allowed to graze along the way, and if they had plenty to eat, and fresh water, the tough Longhorns would gain weight on the drive and fetch a better price at the end.

It was not work for the faint-hearted. Cowboys were almost permanently in the saddle. They slept outdoors, were constantly on the lookout for danger, and endured extremes of weather and hardship. There was the ever-present peril of attack from Indians, who often resented them crossing their territory. Some Indians demanded tribute, which they mostly took in horses – they weren't that interested in the cattle.

The mere mention of the word 'stompede' (as it's pronounced in Texas) was enough to strike terror into the drovers' hearts. Like much else to do with ranching, 'stampede' is from the Spanish – *estampida*. The Greeks called it 'panic terror', and in one old cattleman's phrase, 'It's one jump to their feet and another jump to hell.' A stampede was 'the personification of instantaneousness'. The herd did not ease themselves to their feet like domestic cattle. The jump from lying down to being on their feet was 'quicker than a cat can wink its eye'. The impulse to stampede passed through the animals as though every single animal in a herd of say 3,500 had been wired together and a switch clicked on. No other breed of cattle in the world was as disposed to stampede as the Longhorn. In an instant, a resting herd would become a solid wave of blind unstoppable panic.

The least thing could spook a herd and set them running. A stray dog sneaking up to a sleeping cow; a cough, a human sneeze; the snapping of a twig, the click of a latch, the smell of a wolf; even the end of a rainbow almost touching the leading cows in a herd. Sometimes, if cattle were bedded down for the night too close together, one flicking its tail in the face of another would cause it to jump up with a bawl and set the whole herd off. There was always a cause, but sometimes it was hard for the trail men to know what it was. Often, with their acute sense of smell, the beasts would scent something on the breeze that the men could not: the smell of cooking wafting on the wind from another trail outfit; the sharp smell of a coyote cub approaching the camp, or of an Indian, because Indians often tried to stampede a herd in order to try and pick off scattered animals. One of their favourite tricks was to burn a sack of buffalo hair on the windward side of a herd to spook it.

It was said that if a herd could be driven for two weeks without a stampede, the danger had passed. Being gathered up from their wide-open ranges, crammed together with a horde of other cattle and driven through strange and frightening surroundings by the most fearsome threat to their freedom that the primitive cattle knew, it was hardly surprising that they were prone to panic. On their ancestral range they would have known every rock and bush and tree for miles around, but snatched from the security of the familiar, they could easily be scared.

The commonest and most terrifying cause of stampedes was a thunder storm, especially at night, when the worst ones occurred. People in the west don't live out in the weather any more, as the Texas trail men did – or, nearer to home, the Scottish and Welsh drovers. A cattle trail could take nine months or more, following the North Star from the Gulf of Mexico, where the salt winds blew fresh, to the drifting snows of the High Plains. Wherever they were, cowboys usually slept outside, without tents or other cover, and moonless nights on the open range were darker and lonelier than anybody living in the modern world of electricity, flashlights and well-lit, comfortable houses can imagine. When a man could not see his hand in front of his face, or his sleeping companions next to him; alone in the vast blackness of a prairie night, with not even a tree for shelter and a couple of thousand semi-feral cattle lying somewhere out in the inky gloom, he relied on whatever resources of fortitude he could summon. But nothing could prepare a cowboy for his first stampede.

The speed of stampeding cattle was astonishing. They stretched out low to the ground and ran 'as if the devil was in them'. If the leaders were stopped by a bolt of lightning or a thunder clap, the cattle following them would be unable to slow down or turn and would plunge over them, trampling the animals on the bottom to death. Some would try to wheel round, others darting this way and that, and from the terrible melee would come a horrifying roaring and moaning, adding to the terror of the storm. While they were running, the herd made no sound, but once the stampede was checked, they would set up a tremendous bawling and lowing, mothers calling for their calves, or lost companions.

Some storms were so appalling, the herd would be paralysed beyond stampeding. In July 1878, a terrible tempest blew up on the plain out from Ogallala, Nebraska. It started with flash lightning, then forked lightning, followed by chain lightning, followed by 'the most peculiar blue lightning, all playing close at hand'. After this came ball lightning, rolling along the ground and around the men and their cattle. Then spark lightning and lightning that settled on the men and cattle 'like a fog'. The air smelled of sulphur, and became so hot that the men thought they would be burned up. The cattle were milling about and moaning as if in distress, but made no move to stampede.

Once the cattle got wet, the lightning would play along the curves of their horns; it would dart round the bridles and bits of the horses and the rim of the cowboys' hats. If a cowboy 'popped his quirt' – cracked the little whip with two leather thongs on the end that he carried to encourage the cattle – it would emit sparks, and even the swishing of a cow's tail would bring forth electrical flashes. 'Snakes of fire' ran along the backs of cattle and up the horses' manes. Lightning sometimes raced around the rims of the wheels on the chuck wagon. Men would discard their pistols and knives and spurs, believing it would protect them from a strike. Mexican *vaqueros* wore a ball of beeswax inside the high crown of their sombreros because the substance would not conduct electricity. All profanity ceased; like actors never uttering the word 'Macbeth', cowboys would never risk taking the Lord's name in vain in a storm.

Charles Goodnight, one of the nineteenth century's most celebrated cattlemen, after whom the famous Goodnight cattle trail was named, described the heat generated by rampaging cattle as almost enough to blister the faces of the men on the leeward side of the herd. The visages of men who stayed with a stampede during an electrical storm were sometimes burned 'a brimstone blue'. That is, if they survived the lightning. Many more deaths were caused by being struck by lightning than by stampeding cattle. Cattle tended to run round a man on the ground, even on black nights, and it was unusual for them to trample anyone unless they were so spooked that the press of animals behind them prevented the leaders from dividing around him. A stampede that ran for 25 miles could be scattered over 200 square miles, taking days to round up and re-form.

Texas cattlemen admired the Longhorn for its indomitable spirit and resilience under harsh conditions, particularly trail driving, but were ever alert to the wildness that lay just beneath the skin. A steer was described as 'gentled' if it would not flee from or attack a man on horseback, but it did not mean it was tame. The Longhorn's sheer will to live, its vitality and hardiness were legendary. At a slaughterhouse and packing plant in Kansas in 1870, steers were being killed by pushing them down a chute, where a man with a long sharp lance would strike them behind the horns. They would then drop to their knees and fall through a trapdoor to be dragged to the skinning beds. One big steer was lanced and fell through the trapdoor, but when the knife was at his throat, he leapt to his feet, ran towards the daylight coming through a door, and jumped a storey and a half to the ground. He then swam a quarter of a mile across the Missouri river to a sand bar, where he 'shook himself and turned his head to the shore – at bay'. On another occasion, a big brindle steer made a dash for his life and liberty from a pen in a hide and tallow factory at Fulton on the Texas coast, from where he swam 12 miles across the bay to Lamar.

The Longhorn was seldom troubled by maladies. They had no herd diseases, tuberculosis was unheard of and miscarriage, or spontaneous abortion, unknown. They even knew how to avoid the clouds of mosquitoes that could obscure the sunlight down on the Gulf Coast and torment them to madness. The cattle would drift to the open shore in vast numbers and swim far out into the sea, where they would spend the night and part of the day, riding the waves, with only their heads and shoulders clear of the water.

Their adaptability to climate, terrain and vegetation was continental. They ranged from the tropics of Mexico, far below the southern limit of buffalo migration, to the blizzard-swept prairies of the Dakotas, into Montana and on to the plains of Alberta. They crossed deserts to get to the sweet valley grazings in California; they survived in the bayous of the Mississippi delta and the marshes of Louisiana and the infertile pinewoods of east Texas, where the cattle ticks 'sucked the very life blood from the time the calves were born' and were sometimes so thick on them that they 'covered up the cattle just like shingles on a roof'. When there was nothing to eat, cows could sometimes be seen standing on their hind legs browsing like deer, hooking down twigs and leaves from nine or ten feet up in the trees.

In a drought, when water became more expensive than hay, Longhorns could live for months on prickly pear – a kind of cactus – without drinking. Pear was both a curse and a salvation for the Texas rancher. When there was plenty to eat and drink it was a dreadful nuisance, colonizing huge areas with impenetrable thickets of spiny growth that resisted all efforts to eradicate it. But in a drought it became the salvation of many a herd of cows. Its succulent stems or pads (*nopales* in Spanish) held reserves of water and nutrients and its beautiful flowers presaged a sweet deep-red fruit called a tuna. Every part of the plant was edible, so long as the lethal spines that covered it to protect it from browsing animals could be overcome. Some animals eat pear if there's nothing else available, but they suffer for it, because the spines embed themselves in the soft tissues of their lips and tongues, where they fester into painful pockets of pus and prevent the animal from eating. Once the spines are embedded, there's a good chance the animal will die. Uniquely, Longhorns could chew the whole pad, flesh and spines, and it was unusual for them to be affected.

Ranchers whose cattle were not as tough as the Longhorn developed a way of burning off the spines from the pear pads without damaging the *nopales*, which would not easily burn because they were full of water. At first they used a torch made from rags impregnated with 'coal oil' – kerosene – on the end of a long green pole, which they rubbed against the pads. It was laborious work to provide enough sustenance for a herd of cattle. Later, they used pressurized petrol blow-torches on long handles, called pear-burners, and which shot out a two-foot-long flame at 2,500 °F. It was hot, dangerous work. The operator had to wear heavy clothing for protection against the spines and the wasps that colonized the pear thickets, and boots to protect him from the rattlesnakes, copperheads, centipedes and scorpions that would inevitably be enraged by his work. An effective operator with a flame-thrower could remove the spines from an acre of pear in a day. This would be enough to feed a reasonable-sized herd of cattle. If he repeated the process every day for a few weeks, by the time he had burned the last acre, the first would have regenerated with new succulent *nopales* and he could start the whole process again. This saved many a herd of cattle from starvation.

At first, when the Longhorns took over the plains after the buffalo were slaughtered, they roamed across vast acreages of unfenced land. In a hard time there was nothing to stop them drifting up to 300 miles south, in migrations comparable to those of the buffalo that had preceded them. In a bad blizzard on the plains of the west, vast herds of cattle would be driven south before the snow, moving on until they came to shelter or the blizzard abated.

Fortunes were made and lost during the two or three decades of cattle driving. A steer selling at the railhead in the north for $50 left a profit of around $44, after deducting the purchase price and trail expenses of about $1 a head. The average herd making the journey from Texas was around 2,000 strong, but there were much larger ones, up to 15,000. A man could get rich from just one cattle drive. One of these was Captain Charles Schreiner, whose family had arrived in Texas from Alsace in 1852. After the Civil War, in which he served as a private in the Confederate army, he returned to San Antonio and began trading and cattle driving. He made enough money to buy 27,000 acres and start the Y.O. Ranch,* which in true Texan fashion grew to over 600,000 acres by 1900. Exploiting these huge herds set Texas back on its feet after the Civil War and served as an hors d'oeuvre to the later boom that came with the discovery of oil.

The Longhorn's legendary resistance to the effects of ticks ought to have given it an advantage over the new breeds, but it turned out to be yet another reason why it was driven to the brink of extinction. Over centuries of harsh natural selection Longhorns had developed immunity to tick fever, a disease that was fatal to about 80 per cent of the European cattle that caught it. Some of the cattlemen noticed that if the cows drank sulphur water the ticks would drop off them. Before the open ranges were fenced, when cattle could travel almost as far as they wanted, the herds would go 20 or 30 miles to drink their fill from particular sulphur springs, passing by fresh water to get to the minerals. It was observed in the eighteenth century that buffaloes would make an annual migration of over 200 miles from their usual range, across the Alleghenies, to drink from and wallow in

* The Y.O. brand was first used in the 1840s by Youngs O. Coleman of the Fulton Family Ranching Empire near Rockport.

salt springs in Pennsylvania. There are many accounts of Longhorns making similar pilgrimages to satisfy an annual craving for minerals, particularly salt.

As the Longhorn herds were driven north to the railheads, they carried with them ticks that dropped off and shed their eggs, which hatched into infected larvae and attached themselves to the settlers' cattle. Farmers noticed that their cattle got tick fever after the Longhorns had passed through and assumed that the Longhorns were spreading some disease. As the settlers consolidated their hold on their sections of the prairie and, crucially, barbed wire became available, they fenced off the trails and refused permission to the cowboys to cross their land, both as a response to their fear of disease and also to deny the cowboys a market for the Longhorns, which were competing with their own cattle.

In a typically American response to the threat to the softer European cattle, various experts from the newly formed USDA advocated eradicating the Longhorns altogether. Not only were they spreading disease, but even worse in the eyes of the authorities, the Longhorns themselves remained remarkably healthy. It was another example, along with eradicating the buffalo to starve out the Plains Indians and spraying defoliant in Vietnam, of the kind of scorched-earth policy that the US authorities have favoured from time to time. Such was the hostility to the Longhorn that by the turn of the twentieth century it was in real danger of going the way of the buffalo.

By 1906, when the Longhorns had largely disappeared from their former ranges, tick fever was still endemic throughout 14 southern states and was believed to be limiting the introduction of European breeds of cattle. It was considered necessary to eliminate the ticks by state and federal eradication programmes. For 35 years, efforts were made to try to remove them from the eastern seaboard to the Texas–Mexico border. By the 1940s, a tick-free permanent quarantine zone had been created along the international boundary from Del Rio to the mouth of the Rio Grande, in an effort to prevent ticks crossing into the US from Mexico, where infected ticks remained.

But it proved impossible to eradicate ticks. By the 1970s, they had returned to Texas, where they have proliferated and become a serious problem again. The official USDA response is to quarantine cattle to try to prevent ticks from spreading. But decades of changed land use, encroachment of brush and increasing wildlife have brought them back. By 1 February 2017, more than 500,000 acres in Texas, apart from the permanent quarantine zone, were under various quarantine measures managed by the Texas Animal Health Commission and USDA in collaboration with farmers and wildlife agencies. Quarantine is designed to prevent the movement of animals harbouring ticks and allow the authorities to treat animals that act as hosts. But it is impossible to prevent the movement of wildlife or to treat or remove every host species.

There are two closely related tick species that transmit tick fever: the cattle tick and the southern cattle tick. The veterinary name for Texas cattle tick fever is bovine babesiosis; it is caused by a parasite that breaks down the red blood cells, resulting in anaemia, fever and eventually death. The cattle tick is from the Mediterranean, and the southern cattle tick originated in India. They were both imported into the Americas with cattle from the Old World, and both adapted to the climate of the US, Mexico and Central and South America, where they found an abundance of hosts.

USDA appears to have learned nothing from the failed attempt to eradicate the ticks in the first half of the last century, or the slaughter of the Longhorns at the end of the nineteenth. Animals (and humans) develop immunity to tick-borne diseases. The Longhorn was a good example. Yet instead of working with the cattle to breed a population unaffected by ticks, they tried to eliminate the very type of cattle that would have solved the problem and embarked on a futile campaign of eradication using chemicals, quarantine and slaughter. Recently they have tried genetic modification to create ticks that will be infertile and unable to reproduce themselves. This is unlikely to be any more successful than the other methods. There is something in the psyche of Western scientists, particularly in America, that will not accept that it is better to work with nature rather than try to conquer it. The same attitude embraces pasteurization and genetically modified crops.

Just as the 60 million American buffalo had been ruthlessly eliminated in less than a century with concern for nothing but the money to be made, so the 10 million indigenous Longhorns were seen as a free resource to be pillaged and almost eradicated in the rush for profit. But the Longhorn proved tougher, even than the buffalo, and was not easily dispatched. During the twentieth century it was kept going by a few Texas ranchers, and in the last few decades has made a notable comeback. Its devotees value its extraordinary resistance to drought and disease, its ease of calving small, vital calves, even when mated with bulls from bigger breeds, and its self-reliance and fiercely protective mothering, particularly in the face of danger. Above all, the breed has retained its considerable capacity to convert indifferent herbage into lean meat without artificial feeding or other inputs.

CHAPTER 17

'Rascals with horns goin' straight out'

P RECISELY AT 11.30 a.m. and 4 p.m. each day (weather permitting), a solemn procession of eight cowboys, booted, hatted, bewhiskered and got up as if they were still on the trail, drive a herd of fifteen pampered Longhorns (one for each decade of Fort Worth's history) along the street from their accommodation in one part of the Fort Worth Stockyards, past the Cowtown Coliseum (where the rodeo takes place) and into a large corral, where they pose with their minders for photographs. Then they convey them back to their pens to await the next drive. The docile animals plod along in the fierce summer heat, their massive horns nodding from side to side in time with their steps. The parade resembles a Catholic feast-day procession through the streets of some Spanish town, accompanied by a silence similar to the awe that would greet the parading of the relics of the local saint.

Until you see these cattle in the flesh, it is hard to appreciate how unusual they are, from the spread of their horns – up to eight or nine feet from tip to tip – to the way they quickly learn to negotiate narrow gates and alleyways by turning their heads sideways and looking forward out of one eye, even deftly reversing into narrow spaces so that their bodies are inside while their head and horns remain outside, to their unique multicoloured coats, with no two animals having the same combination of colours.

In the early 1960s, the Texas Longhorn Society began to register the breed. Longhorn cows then had horn spreads of between two and three feet. As stock was selected for size and muscle and also horn length, the average spread from tip to tip increased by about a foot during the seventies, and continued growing into recent times. With careful selection for horn spread, the average has increased by about an inch a year for the last 40 years. Coupled with enhanced carcase quality, the breed has been improved considerably without any out-crossing or loss of hardiness or ease of calving. It also appears to have lost a good deal of its fierceness. Whether this is the result of breeding or handling I cannot say.

There is also the delightful paradox that in a new country, this is a very old breed, whose ancestry can be traced back over 500 years to a precise year at the end of the fifteenth century: 1492, when Columbus sailed the ocean blue. Then, in less than a century, they went from being numbered in their millions to being almost wiped from the face of the earth. In recent decades they have acquired a group of passionate devotees, and a cult has arisen around them. They have become the emblem of the state of Texas, and of a city, Fort Worth, whose tourism is partly sustained by the myth of the Texas cowboy; and they are celebrated, even revered, in the twice-daily ritual that has the scent of the religious about it.

This is living history with the difference that all the players have worked as proper ranch hands in the past and know their business. 'It's an awful lot easier than doing it for real,' said one of the grizzled horsemen, every one of whom has some form of facial hair, impressive moustaches and goatee beards being very popular. The drive passes the Fort Worth Livestock Exchange, once known as the Wall Street of the West because of the enormous numbers of cattle (and other livestock) that came here to be turned into money during the boom years. The drive and the cattle are managed for the Fort Worth Convention and Visitors Bureau by a dynamic Polish girl called Kristin Jaworski. She has absorbed fully the Texas enthusiasm and clearly loves what she does with 'the world's only twice-daily cattle drive'. Each beast has its own calling card, with a portrait photo on one side and details of its date and place of birth on the other, together with some snippets of information about its life. These cards are put together into a pack sold to tourists.

Fort Worth, established in 1849 on a bluff overlooking the West and Clear forks of the Trinity river, was the last halt on the Chisholm Trail before it entered Indian country. In 1876, the Texas and Pacific railroad arrived in the town and turned Fort Worth into a terminus for livestock, with their products shipped out all over the country. By 1900, the railway was bringing in vast numbers of animals for sale and slaughter in the stockyards, turning Fort Worth into a hub for an enormous selling, slaughtering and processing industry, from which it got the nickname Cowtown. In the last century or so, more than 160 million head of cattle, sheep and pigs have been sold through the Fort Worth Stockyards, which at its height held the biggest sales in the south-west of the US.

For a few years, the Chisholm Trail was one of the routes by which the cowboy drovers, after the Civil War, brought their

Longhorns from Texas to the railhead in Kansas, from where they were sent eastward to feed the terrific demand for beef, tallow and hides from the mushrooming cities. The droving only lasted a couple of decades, but that hasn't prevented it from passing into the myth of the western cowboy, more romance than fact. Hundreds of thousands of cattle that had been rounded up across the plains and tracts of sage brush of southern Texas between the Rio Grande and the Brazos river were driven north to converge at Red River Station, where they crossed the Red River on the border between Texas and Oklahoma. One major branch of the trail passed through Fort Worth.

The trail is named after Jesse Chisholm, born in Tennessee, son of a Scots father and Cherokee mother, who as a frontier trader with the Indian tribes established the route with easy gradients and river crossings to carry the heavy wagons that transported his goods. Chisholm was among the early pioneers who moved west into what is now the state of Arkansas. He acted as a go-between for the Indians and the American settlers, using his knowledge of both cultures and an alleged ability to speak 14 different Indian dialects as well as English.

In 1867, the year before Chisholm died, Joseph G. McCoy, a cattle buyer from Illinois, persuaded the Kansas Pacific Railway to lay track to Abilene in Kansas. He built holding pens and loading facilities at the railhead to attract the Texas cattlemen to his new facility. In that year alone, some 35,000 head of cattle came along the Chisholm Trail to Abilene, consigned to McCoy's stockyard, which became the largest west of Kansas City. Within four years, 600,000 animals had passed through his yards. During its short zenith, an estimated five million head of Longhorn cattle travelled over the Chisholm Trail, their hooves churning up a track that was in some places 400 yards wide and over time, eroded by wind and water, indented the plains it crossed; in places the trail is still evident today.

Homesteaders settled and fenced the land through which the trail ran, and brought their own breeds of European cattle: Aberdeen Angus, Hereford and Shorthorn. The pioneers had little time for the indomitable Longhorn and were keen to fence against it, partly to demarcate their square-mile sections (640 acres) and partly to prevent their imported cattle from being 'contaminated' by the feral Longhorn. The drovers were forced further west, into eastern Nebraska, and by 1875, with the coming of the Union Pacific railroad, Ogallala in Nebraska became the main shipping point. The route to it was known as the Texas Trail, which superseded the Chisholm Trail as the main route north through Indian territory. Between 1867 and 1884, when the last drive took place, millions of cattle and mustangs went up the trail. By 1885, the vast stocks of Longhorns had been plundered, and then the new railroads replaced the cattle trails and the short heyday of the cowboy had passed into American myth.

The Longhorn breed was brought to public attention when the University of Texas adopted a Longhorn bull, Bevo, as their mascot in 1917. But it was not until 1927, when the last of the breed was almost gone, that a few private landowners and enthusiasts from the US Forest Service gathered together a small herd from the animals that were left, keeping them in the Wichita Mountains Wildlife Refuge in Lawton, Oklahoma.

Then, a few years later, J. Frank Dobie (who wrote *The Longhorns*), with others interested in the breed's salvation, began to collect a few animals in small herds, mostly as curiosities. Their sterling qualities of longevity, disease resistance and astonishing thriftiness on moderate grazing then attracted commercial interest as beef stock. That, coupled with a sentimental attachment to their place in Texas history, has ensured the breed's resurgence, although it has to be admitted, keeping Longhorns is largely a rich man's hobby. They can now fetch enormous prices: $40,000 at auction is not unusual, and in March 2017 at a sale in Fort Worth, a record $380,000 was paid for the cow 3S Danica with her heifer calf at foot. There is a small yet growing demand from health-conscious consumers for the Longhorn's lean beef.

With all this in mind, I arranged to visit a couple of herds in Texas. I was put in touch with Rebecca Wampler, who keeps Longhorns near Mineola and has donated animals to the Fort Worth Stockyards for the daily cattle drive (or plod might be more accurate). She was helpfulness itself, and arranged for me to go to the Longhorn Cattle Breeders' Association 'Mid-Year Blowout Sale' (nobody could accuse Texans of hiding their light under a bushel) on 17 June 2017, at West, north of Waco in Texas.

The West Auction Barn stands prominently on an open site beside the southbound Interstate 25, convenient for transport but not a prepossessing place, with the constant roar of heavy traffic. Viewing was held the evening before the sale so buyers could inspect the cattle being brought in to be ticketed and settled in their pens for the night. There were 92 lots to be sold next day. Compared to cattle auctions in Britain, the West Auction Barn is a primitive affair, with gravel rather than concrete on the floor and rickety rusty metal fencing round the pens that lie behind the crescent half-ring through which the animals come to be sold. But it works.

The evening was very hot – about 100° F – and most people were sweating freely. Only the cattle showed no obvious discomfort in the heat. Huge pickups attached to goose-necked trailers that looked as if they were feeding from the back of them drove right into the holding pens behind the auction building and let down their tailgates to release their contents. Russell Fairchild, the sale organizer, bustled about, penning cattle, ticketing, checking paperwork and taking responsibility for feeding and watering the animals. He was very good at it. He even found time to introduce me to everyone who was anybody, and once they'd determined I wasn't Australian – which most of them thought I was – everyone had something to say about the English and England.

I was leaning against a gate in an alleyway when a great shout went up and a cow with horns almost too wide to fit between the pens came bowling towards me with her head slightly inclined to one side, one horn pointing straight at me like a javelin. 'Watch out for this one!' shouted Russell. 'She's crazier 'n a bed bug!' and everybody jumped out of the way, or climbed the metal bars of a pen, to let the roan cow with 80-inch TTT horns past. (TTT is a characteristic American acronym meaning 'tip to tip'. Why they don't just say tip to tip is one of those impenetrable mysteries of the American psyche.)

Kurt and Glenda Twining of Silver T Ranch have a commercial herd of Longhorns, but Kurt freely admits he couldn't afford to keep them if he didn't have another source of income. Lot 20 is one of his heifers. She has elegant black-tipped horns, already impressively long, and he's certain they'll grow much longer because her mother had 80-inch horns, as did her father. A seven-foot spread of horn is not unusual, he tells me. It is hard to see how they keep hold of such huge horns, which are bigger than the top of the head they grow from. There must be a lot of leverage at the root if the tip gets caught in something.

Longhorn beef is becoming saleable again, even valuable, after almost a century of not being wanted. It makes about $7 a pound (a far cry from the 10 cents it made at the Kansas railheads) and is bought by people who value the fact that it is grass-fed, lean, and free from chemical stimulation and corn or soya feeding. This is the nearest you get in the arid parts of the US to living off the terroir. There are newer cattle breeds on the prairies, such as the Angus, which are kept in a similarly natural way, but they are nowhere near as tough as the Longhorn.

However, most Longhorns are now being bred for horn length rather than anything to do with beefing qualities. Every beast that struggled with its impressive horns into the ring had the width of spread announced and the measurement recorded in the sale catalogue. A decent set of horns (attached to a head) now fetches around $350 – about a third of the carcase value. Those with particularly impressive antlers were the object of extravagantly expressed admiration by Russell Fairchild and the auctioneer.

It takes nourishment to grow horn. Steers – bulls castrated as calves – grow the longest horns. Entire bulls and cows do not develop the same length. It was believed among Mexican cattle keepers that a cow needed horns to thrive because it needed *un lugar para sangrar* – a place to bleed. Texans picked up an echo of this when they thought that if an animal was not doing well it was suffering from 'hollow horn'. To remedy this they would either bore a small hole in each horn, or saw them off short, with much loss of blood.

From this arose the saying, when referring to a person who seemed particularly listless or dopey, that he needed 'boring for the holler horn'.

Horns grow until an animal is 10 to 15 years old. Straight at first, and then with twists or wrinkles as it reaches maturity. 'Mossy-horn' came to describe older, rougher cattle with lots of wrinkles on their horns, as well as any veteran cowman with age and experience, in contrast with a greenhorn, a beginner. It was also used to refer to horns that were twisted about with low-hanging Spanish moss from bushes and trees.

Practically all the cattle for sale at West had wide horns that grew out and up, some with black points, others with uniform colour the full length. But hardly any of them had horns that were set for hooking or goring. They had what Dobie calls an 'exhibition spread' – more for show than utility. Nothing can now compare with the massed horns that 3,000 Longhorns presented. As the herd grazed in tall buffalo grass, their horns could appear disembodied, weaving and bobbing in the air, disconnected from anything.

On the long trails, cattle that had been reared on and adapted to dry prairies and wide plains had to swim across every body of water that crossed their way. There are few rivers in Europe that compare with the mighty watercourses in America. The old trail bosses describe crossing rivers in flood: apart from the glistening sea of horns, the cattle would disappear beneath the swirling brown water, the herd undulating and surging in a long wavering S as they fought the current, swimming for their lives to the opposite bank. This was 'something that not the Colorado, the San Gabriel, the Brazos, the Trinity, the Canadian, the Arkansas, the Platte or the Yellowstone ever surpassed in wonder'.

Horns were not just for decoration or defence. They were supremely useful to pioneering people for myriad uses: hand-cut buttons, cups and spoons, pegs and racks, pieces of furniture such as hat-stands and tables. The cattle kings of the 1880s and 90s had a passion for chairs made of horns. Horned skulls were used as signposts. In some desert places water was carried in horns. A long blowing horn hung at a crossing on the riverbank to summon the ferryman. On plantations it called field hands to dinner.

In *Far Away and Long Ago*, W. H. Hudson describes fences seven, eight and nine feet high built of hundreds and thousands of horned skulls enclosing fine houses in Buenos Aires. Some of the older walls were festooned with creeper, wild flowers and green grass growing from the cavities in the bones; they 'had a strangely picturesque but somewhat uncanny appearance'. During the mission days in California, just as householders used to cap their walls with shards of broken glass to deter intruders before they were made liable for the welfare of burglars, so people used horned cattle skulls to top the adobe walls of their corrals to make a palisade against horse thieves.

From the *conquistadores* to the gold hunters of California, prospectors used a great spoon made from the horn of a cow split lengthways, softened by heating and fashioned into a kind of scoop, rather like the shape of a wide Indian canoe. As the deposit of crushed stone or gravel was swirled round in water, the lighter material came to the edge and could be washed away, leaving the flakes of gold in the hollow.

Sets of horns adorned bars, saloons, hotels, public buildings and houses all over Texas. Most banks had at least one bovine head on the wall of their lobby, in recognition of their main source of revenue. Businesses with even a tenuous connection with cattle or ranching used the emblem of a head of horns on their writing paper and advertising. It was the favourite design on the handle of pistols, engraved in bone or horn. Leddy's bespoke boot makers in Fort Worth, and nearly every other shop and bar for miles around, has a pair of cow horns adorning the wall. The head and horns symbolizes strength, power, the freedom of the land in the Big Country, and is also a memorial to the formidable animal that thrived in a hostile land and within a few short decades was sacrificed for the fortunes of the State of Texas.

To some extent, terroir affects the length of horn, but it seems that inheritance has the decisive influence. When J. Frank Dobie and Charles Goodnight were engaged in collecting the remnant of the Longhorns for preservation in the Wichita Mountains Wildlife Refuge in 1927, Goodnight believed that under such conditions their horns would never grow to the impressive spread of the original range cattle. He thought they would become shorter and thicker and their bodies more compact, because 'no power on earth will defeat nature'. By 1940, Goodnight's prophecy seemed correct, because there were few spreads wider than four feet. He associated 'mighty horns, like the hoarse howl of the lobo, the wide wheeling of the eagle, and the great silence on the grass, to be a natural part of the freedom, the wildness and the self-sufficiency of life belonging to the unfenced world ... the crown of the open range'.

But judging by the horns on the cattle offered for sale in the ring at West Auction Barn, time and breeding have proved him wrong.

The Texas Longhorn Blowout Sale was 'hosted' by two breeders who provided most of the cattle for sale: Mike

MacLeod of Split Rock Cedar Ranch and Dr Zech Dameron of Clear Creek Pecan Plantation.

About a hundred buyers assembled on the raked seats above the auction ring, which was a semicircle, with the rostrum on the flat side. Russell Fairchild's fruity Texan drawl introduced the key people: Keith Presley, auctioneer; the ringmen; and the 'pretty good-looking old girl here behind me' who owned the mart. Russell's father was in the crowd, and he had him stand up and introduce himself. Then he got Mike MacLeod onto the rostrum to lead the prayers.

Mike was remarkably good at extemporizing on the theme that we should thank God for creating the cattle we were about to see passing through the ring. The prayers covered everything we had to be grateful for and thanked God for his generosity. Then he broadened out his theme. 'Last year, as you all know, I lost my dear wife.' He paused, and a general murmuring of regret for his loss came from the crowd. 'But the Lord moves in mysterious ways! Praise the Lord! So here's mah new wife, Teresa. She's been a blessin' from the Lord!' A few 'Amens' and a general murmuring of appreciation for the Lord's goodness came from the crowd. Mike had Teresa stand up and beam round the auditorium at the buyers, acknowledging the clapping and generally basking in the limelight.

Then Mr Presley started selling. He rapped his gavel on the rostrum to get attention and began each lot with 'Right, here's the deal!' singing his auctioneer's patter, rolling his Rs, keeping going until he had to breathe. The first lot was a 'bred' (in-calf) heifer from the same Mike MacLeod. (She had been 'exposed' to his bull, Valentino – maybe a clue to old Mike's romantic nature that had attracted Teresa.) Presley was very good at keeping up the excitement, with the help of his characteristic Texan enthusiasm – and the ringmen. We don't have ringmen in Britain, or at least not specifically employed as such. Now and again auctioneers will take turns to sell, with the one not selling watching for bids and pointing out anyone the auctioneer might have missed. But the American system employs qualified auctioneers who stand on either side of the ring, scanning the crowd for bidders. When somebody bids or appears to bid, they point at them and shout, 'Hup!' or 'Yep!' with great gusto, whipping up excitement. The shout of the ringmen is the Texas yell. It was born 150 years ago, when cowboys were overcome by the thrill of roping cattle at night or riding down a maverick. It's a combination of the Comanche war whoop and the wild shout of the Johnny Rebs as they charged the Billy Yankee lines during the Civil War – which is not forgotten down here.

All the important men wear white cowboy hats, white shirts, jeans and fancy tooled boots. Their women wear shorts or short skirts with cowgirl boots and are the thinnest, most attractive I've yet seen in this land of obesity. Their menfolk must have a lot of money, because the richer you are in America, the thinner you can afford to be.

'What kind of mama is she?' shouts one potential bidder as the next cow twists its neck to negotiate its vast horns through the gate into the ring. The cow seems to know instinctively where the tips of her horns are and avoids getting them jammed in the metalwork of the pens. These horns are not meant to be worn by cattle that live inside or in feedlots.

'She ain't mean! She's a lady,' declares the auctioneer, giving away nothing with his ambiguous answer.

'You got to git one of these rascals with horns goin' straight out!' he calls apropos of nothing when the bidding flags a little. He is taking bids from the internet picked up by two women with laptops sitting on a platform above the ring.

'OK, here's the deal! You bid on her, you pay the money, you get the calf!' he tells another bidder who seems to have lost enthusiasm for his purchase. 'It's just a hundred bucks, dude,' he urges another who won't go again. 'Cash or check, we take it all here! It'll be the best thing you've done all day. It only hurts for a little while! I know you want the cow. You'll be proud of what you done when you get her home!'

The buyer looks for approval from the woman sitting with him, and the ringman sees this and shouts, 'Don't worry about asken her! I'm a licensed marriage counsellor and Ah'll make it all right!'

There's more shouting and whooping and then the auctioneer says, 'Guess what! You've just bought the cow! Thirty-seven hundred!' and everybody cheers. 'Thank you, sirrr, for staying with me. Thank you, ma'am.'

After two or three lots, Russell stops the auctioneer and announces that he has 'forgotten to introoce a famous author from the north of England' who is here to find out about Longhorns for his next bestseller. I have to stand up and turn from side to side, thanking the crowd for having me and giving me their Texan welcome.

'He's called Phullupp Walling – dubbya ay ell ell ah enn gee!' And they all clap.

And so the sale goes on, for four hours, with a great buzz created by a combination of the crowd, the auctioneer and the ringmen.

The last lot is a cow called Jamajawea. 'We have loved her, loved her, loved her! She's a great mama. And though she's open, she's a guaranteed breeder. We will guarantee her. We flushed her and got lots of babies from her. Sixteen embryos. That's why she's taken a bit of time to git bred again. Here's the deal! Have I got three thousand? Thirty hundred? Thirty hundred anywhere? Twenty-fahve hundred then?'

'Yep,' barks the ringman and points jubilantly into the crowd. 'Don't let him scare ya, ma'am! You know you want this cow. You just bid. You'll be happy when you get her home! She's a nice big prutty cow!'

The bidding hots up, with the ringmen yelping and whooping back and forth and pointing to bidders on their respective sides of the ring.

'Yep!' from one side.

'Hup!' from the other.

'Yep!'

'Hup!'

'Yep!'

'Hup!'

'Sixty-two fahve!' shouts the auctioneer.

'Hee-haw!' whoops the right-hand ringman.

'Sixty-fahve! I got sixty-fahve,' yells the auctioneer.

'We got seven thousand?' asks the left-hand ringman of the woman he's encouraging to bid.

'Nope. Sixty-seven and a half,' replies the auctioneer.

'We do finance if you like, ma'am! You can't go home with an empty trailer! If they say guaranteed, you'd better take it. I know these guys.'

The bidding edges up. 'Eighty-fahve hundred dollars. Last shout today … Eighty-nine hundred? Make it ninety? Ninety it is! Thank you, sirrr! Make it ninety-one?'

The woman shakes her head.

'Nine thousand dollars! Ladies and gentlemen, Ah just sold a cow for nine thousand dollars!' and he bangs his hammer triumphantly on the rostrum. General whooping and clapping.

These cattle are a long way from the fabled old Longhorns. They might be thrifty and tough, but they are tractable and docile. They patiently negotiate the pens and the ring, and although they could do a good deal of damage with their outrageous horns, if they did, it would only be accidental. I saw only nervousness, not aggression. They do have a kick like a horse, both feet backwards at the same time, but that is only in reaction to something startling them.

Texas used to be called the 'Rawhide State', and it was said that 'What a Texan can't mend with rawhide ain't worth mending.' Rawhide is, as the name suggests, the raw, untanned skin of cattle – originally the Texas Longhorn – and was part of the culture of the country. The earliest Spanish grants of land were measured not in chains but in *riatas*, ropes made of rawhide.

Rawhide has passed into myth, like much of the pioneering life of the nineteenth century. Films, adventure stories and folk tales have turned the very word into a romantic ideal totally at odds with the truth. In a parched land with almost none of the products of the soil that they were used to, the settlers turned to the most versatile and readily available substance to hand: the skin of the almost unlimited cattle with which they shared the country. Texas was 'bound together with rawhide'.

Tanning preserves the skins of animals and makes the resulting leather supple so that things made from tanned leather will remain in the shape they are crafted in. Just think of your shoes or handbag. But rawhide is the untreated hide of the cow, with the meat and fat scraped off. It contracts as it dries and expands when wet and is immensely strong and almost untearable, with the capacity to stretch to enormous lengths. When properly dry, it becomes as hard and durable as wood.

There are many stories about rawhide's almost unbelievable power. Some are true, some exaggerated and some pure fiction, and it is hard to distinguish between them. I suspect this one is in the fiction category. A settler went to a creek to fetch water with his 'lizard' (a sled that carried a barrel and was dragged by a horse). After he had filled the barrel, a storm blew up, and to shelter himself from the driving rain he walked on the lee side of the horse back to his house. At the kitchen door, he turned to see the horse was still attached to a pair of long, thin, stretched-out rawhide traces leading back down to the creek, but no sled.

The settler unhitched his horse and threw the hames (where the traces are attached to the animal) over a tree stump. The rain stopped, the sun came out and the earth started steaming in the fierce sunshine. He settled down in the shade to chew tobacco and wait. At length the sled with the barrel on it came into view, pulled by the contracting hide, inching its way home and eventually stopping right at the tree stump.

Another story, which I would like to be true, is about a haulier travelling with two wagons each drawn by eight oxen. One of the wagons got bogged down in black peat in a creek, and even attaching all the oxen to it failed to extricate it, despite the fact that oxen are good steady pullers, not lungers like mules or horses. Presently a Longhorn range bull bellowing some way off gave the haulier an idea. He sneaked up on the animal, shot him and had his men skin him. While the hide was still warm, fresh and pliable, he soaked it in the creek for an hour or two and cut it round and round into a long strip. He tied one end to a tree on the bank, then stretched the strip of hide tight and fastened the other end to the wagon's drawbar.

The sun shone fiercely all afternoon while the haulier and his men sat in the shade, drinking coffee and smoking their pipes. He took a nap and filled his pipe again and watched the wagon for signs of movement. At length he noticed the spokes of one wheel move slightly. The sun beat down all afternoon, and gradually, almost imperceptibly, the wagon wheel moved again, but only a fraction of a revolution. When the sun went down, the wagon had moved slightly forward. The haulier unfastened the rawhide and soaked it in the creek overnight. Next morning, he fastened the strip to the wagon again and let the sun do its work. The rawhide pulled slowly but inexorably, and by nightfall the wagon had moved another infinitesimal amount. He soaked the rawhide again for the second night, and on the third day attached it again, and during the day it drew the wagon out of the mud and up the bank.

Not all the uses for rawhide were benign. One Mexican who had conceived a dislike for Spaniards used to sew up any he caught in fresh hides and leave them in the sun for the rawhide to dry and slowly crush the unfortunate captive to death. In South America it is said that prisoners were disposed of by tying their limbs to four posts with fresh rawhide. They were exquisitely quartered as effectively as if they had been hitched to horses driven in opposite directions.

This ruthless exploitation has become transmuted by sentimentality into the tourist attraction that is the Fort Worth Stockyards and the adoption of the horns of a Longhorn as the symbol of Fort Worth, and into all the other myths about the cowboy and the Wild West. But when the industrial killing of millions of Longhorns was going on, many places would never have been free of the noise and smell of terrified cattle, their blood, urine and muck running down the streets. It is almost an affront to the memory of the Longhorn's elemental vigour and pride to see the pampered steers plodding down the street in Fort Worth twice a day; too painful to dwell on the cruelties that would have been routinely inflicted on millions of terrified animals by desperate men whose greed drove the Longhorn closer to the brink of extinction than ever the buffalo was.

The Longhorns pushed through the ring at West and those that live in pampered luxury on their modern owners' ranches bear only a superficial similarity to the cattle that once grazed the western ranges. They have little of the old Longhorn about them that Dobie so admired: that 'adamantine strength, aboriginal vitality, Spartan endurance and fierce nobility ... that made one associate them with Roman legions and Sioux warriors'.

CHAPTER 18

The Spanish Fighting Bull

> Bullfighting is the only reality.
>
> Federico García Lorca (1898–1936)

> Any man can face death, but to bring it as close as pos-
> sible while performing certain classic movements and do
> this again and again and again and then deal it out your-
> self with a sword to an animal weighing half a ton which
> you love is more complicated than just facing death. It
> is facing your performance as a creative artist each day
> and your necessity to function as a skilful killer.
>
> Ernest Hemingway (1899–1961)

DOWN THE AGES, mankind has chosen his domestic cattle to be tractable. Docility was as valued as durability. The oxen that plodded the croplands of the world were our willing companions, with never a thought of freedom from their yoke; the uncounted droves from the furthest parts of the kingdom, fording rivers and crossing seas; the patient dairy cow whose milk has sustained us from the dawn of time: in all of these we bred and valued obedience.

But there is one kind of cattle, reared in large numbers across southern Spain, into which exactly the opposite characteristics have been bred; beasts selected over the centuries for their ferocity and intractability. The *toro bravo* is unique to the Iberian peninsula. Where it came from and how it got its character will never be known for sure, but it has existed for many centuries for the single purpose of dying nobly in the arena in a twenty-minute ritual sacrifice, at the hand of a man who risks his life time and time again in the tragic spectacle.

All over Spain, the most ubiquitous image is the Osborne bull, first used in 1957 to advertise the firm's Veterano brandy. The iconic black silhouette could be seen beside nearly every major road in Spain and Portugal until 1994, when the EU prohibited all advertising of alcoholic drinks closer than 150 metres to a road and ordered the image's removal. Following a public outcry, a compromise allowed the images to be retained, but with all words blacked out. Osborne moved them back from the roads to comply with the 150-metre rule, but cleverly enlarged them, with the result that the new silhouettes are even more dramatic than the originals. And when the EU court ruled that the signs had become part of the landscape and had 'aesthetic or cultural significance', the Osborne bull became the symbol of Spain.

My son's wife's family, the Caballeros, live in Seville and have been making sherry on the delta of the River Guadalquivir for many generations. This part of Spain is also fighting bull country. One of the Caballeros married a Miura, a family that has been in the premier league of fighting bull breeders for over 175 years. So I used this connection to get an invitation to Zahariche, the Miura brothers' *ganadería*, near Lora del Río, in Andalusia. Don Antonio Miura was a little hesitant when my son first phoned him, and I wondered if it was because he feared this Englishman would be critical of his work. But that wasn't the case at all, because the Miura family has had a lot to do with Englishmen over the years. Don Antonio's nephew, Eduardo Miura, befriended the Englishman Alexander Fiske-Harrison and taught him enough about bullfighting to get into the ring and kill a bull. (He wrote about it in 2011 in *Into the Arena*.)

The metal gate from the public road into the Miuras' land, held up by three old telegraph poles, is an extreme example of home welding. 'Miura' is emblazoned along the top in rough letters two feet high, fashioned from what look like bits of scaffolding pipe. It is flanked by two bleached bovine skulls complete with horns, reminiscent of the entrance to a ranch in a cowboy film. The huge gate catch can be opened by a rider without dismounting by pushing a long lever that sticks up above the top bar of the gate.

Recently one of the Miura bulls got out onto the public road and so terrorized the stretch of highway that despite his value, the police shot him rather than attempt to recapture him. It took thirty 9 mm bullets to kill him. The reputation of the Miura strain, nicknamed the 'Bulls of Death', has passed into legend. Juan Belmonte, one of the greatest bullfighters in history, said of them that 'no bull ever showed greater offensive and defensive capacity in the face of the bullfighter. All the other bulls I have ever fought could eventually be brought to the point of absolute submission; the Miuras never.'

Ernest Hemingway wrote in *Death in the Afternoon* that 'There are certain strains of bulls in which the ability to learn rapidly in the ring is highly developed. These bulls must be fought and killed as rapidly as possible with the minimum exposure by the man, for they learn more rapidly than the fight ordinarily progresses and become exaggeratedly difficult to work with and kill. Bulls of this sort are the old caste of fighting bulls raised by the sons of Don Eduardo Miura of Sevilla.'

No other strain of bull has killed more men than the Miura. They are the only bulls to have caused a matadors' strike. As the nineteenth century drew to a close, the death toll of leading matadors killed by Miuras grew uncomfortably long. In 1908, 15 top matadors refused to fight at all unless they were paid double to face a Miura. The promoters responded by bringing on young fighters who were aching for a chance to make their names and thus broke the strike. The 15 matadors were forced to capitulate. As if to prove their point, two of them went on to join the illustrious band of matadors killed by a Miura: Pepete in 1862, El Espartero in 1894, Domingo Del Campo in 1900, Félix Guzmán in Mexico in 1943 and, most famously, Manolete in Linares in 1947. That is not to mention the amateurs whose names have not been recorded in the official record, or those who have been disabled or seriously injured.

The injuries are bad enough. The great banderillero (the man who places the barbed coloured sticks – *banderillas*, 'little flags' – into the bull's shoulders), José Antón Galdón, 'El Niño de Belén', lost his hopes of reaching the top as a matador, along with his right eye, to a dangerous bull. The legendary Juan Padilla lost an eye and one side of his jaw when he was gored in the face in September 2011 at Zaragoza. The horn entered his neck and emerged through the eye socket.

The Miura breed began in 1842, when Juan Miura bought 220 Gil de Herrera cows, and 200 José Luis Alvareda cows and bulls. These came from the Gallardo family, who farmed near El Puerto de Santa María, where the Caballeros have a bodega, and where my son and daughter-in-law were married. Other cattle were added over the years from historic Spanish breeds, including Cabrera, Navarra, Veragua and Vistahermosa-Parladé. These have all gone into making the formidable cattle reared at the Miuras' *ganadería* at Finca Zahariche.*

The low buildings at Zahariche are whitewashed against the summer heat and roofed in yellow-brown fireclay tiles weathered into beauty; the woodwork is bright blue and the whole ensemble forms a defensive square around a large courtyard with a narrow gate on one side, characteristic of haciendas all over the Spanish-speaking world. The courtyard is paved with little rounded cobblestones gathered from the land, sorted into sizes and colours and set in complicated pleasing geometric patterns.

In the middle of the courtyard, water trickles from a pump into the hexagonal trough decorated with blue and white ceramic tiles. Various large dogs, chained outside their kennels to rings set in the walls, are curled up sleeping on the cobbles in the December sunshine. Three long-legged grey-speckled Andalou horses are quietly waiting on their tethers, tacked up with high-backed *vaquera* saddles with sheepskin

* Ferruccio Lamborghini, the car manufacturer (who incidentally was born under the sign of Taurus), was so enthralled by the bullfight that he adopted the fighting bull as his marque in 1963 after visiting the Miuras' *ganadería* the year before. The names given to his cars are nearly all bullfighting related: Miura, Espada (the matador's curved steel sword and a synonym for the matador himself), Islero, Jarama, Jalpa, Diabolo, Murcielago, Gallardo, Aventador, Veneno, Huracan, Reventon, Urus, Asterion.

covers, and the traditional stirrups (*estribos*), that look like a cross between a coal scuttle and a garden trug. The whole foot rests on a plate with triangular steel guards welded to each side to protect the ankles from blows and to make it easy for the rider to remove his foot in an emergency. Draped across the front of the saddles are *manta estribera*, traditional shawls woven from undyed white, black and brown wool. These protect the stockman (*vaquero*) from the weather and give him something in which he can wrap his victuals. From each brow band hangs a *mosquera*, a long fringe of leather strips that swish with the horse's movement and fends off flies in the summer.

Now and again one of the waiting horses lifts a hoof and puts it down again emphatically on the cobbles with a hollow clack that echoes round the walls of the still courtyard. Men in riding boots, tweed caps and studded leather chaps emerge from doors. One is carrying a bucket, another a piece of harness; another is stuffing his arms into a brown leather jerkin and striding towards a horse.

Then it dawns on me that these elegant horses are tacked up for us! I haven't ridden a horse in 30 years, and even then it was only a quiet little fell pony. And I can't recall my son, who I've brought along to interpret, ever having ridden, although he might have done since he came to live with the Caballeros. I'm not sure he would have offered to come if he'd realized he'd be expected to ride out amongst fighting bulls. I'm just coming round to the idea of giving it a go, stiffness notwithstanding, so long as somebody gives me a leg up, when a battered little white Suzuki reverses into the yard. Don Eduardo Miura emerges, and with grave courtesy introduces himself and invites us to get in.

We set off along a rough track around the back of the finca, across a tiny stream and out into the fields. The Miuras employ between 15 and 20 people on their 1,500 acres to look after about 600 cattle and breeding horses and mules. A few of the men are sitting with their backs to a wall, sheltering from the wind, eating their mid-morning bait, as we bump past.

There is little difference between the way the Miuras manage their cattle and the management of an extensively grazed beef herd anywhere else in the world. The sexes and year groups and pregnant and nursing mothers are reared separately. But the crucial difference, I learn almost immediately, is that these cattle are mental. We stop in a field of two-year-old bulls (*erales*) so I can get a photograph. I go to open the door to step outside for a good shot of an animal that is standing on a little hillock about 150 yards away. Clods of turf hang from his horns and he is languidly pawing the ground and flicking up soil with his front hooves onto his back. Don Eduardo grips my forearm, 'Don't get out!' He shakes his head and wags his finger. 'It's not safe.'

Bearing in mind that the biggest thing about the rather scrawny beast is its horns, and it is quite a long way off, I judge that I could open the door and get a shot or two before the little bugger could reach us. I persist. 'I only want to get out for a moment. I won't move from the vehicle. It's just to rest on the open door to get a better picture.'

He wags his forefinger at me, shakes his head and says emphatically 'No! *Muy peligroso!*'

That is when I notice that the beast has stopped pawing the earth and is staring at us like a thug looking for trouble on a Saturday night out. It starts to move towards us, slowly at first, gradually gathering speed. Don Eduardo puts the Suzuki into gear and accelerates away, bouncing over the rutted ground with the little bull gamely following until it seems to lose interest, content that it has seen us off and satisfied some atavistic sense of honour.

'What would it have done?'

'Probably attacked the vehicle and tried to turn it over with its, er ... *cuerna*.'

'Horns,' my son helpfully interjects.

I didn't know then that it is forbidden for a fighting bull to see a man on foot (as opposed to mounted) before it goes into the ring. If it were to gain early familiarity with its two-legged enemy, it would be a cleverer and much more dangerous opponent when it enters the arena for the fight of its life. In fact, it would probably be unfightable. That is why a bull is only allowed to fight once. Afterwards, it is usually killed; in rare cases, its life is spared with an indulgence (*indulto*) granted by the president of the *corrida* on a petition from the crowd. It is then treated for its wounds and goes back to its native ranch to live out its days as a stock bull – often for 20 years or more.

We bounce along from field to field, with me hopping out to open the gates, affecting nonchalance and keeping a weather eye out for cattle. The *toro bravo* is not, as I expected, uniformly dark-coloured, but ranges across most bovine colours from creamy white through red, brown, blue-roan, speckled, to glossy black. The horns in both sexes are massive, dark-tipped and very sharp. On the younger cattle they are out of proportion because they grow faster (about 1 cm a month) than the slower-developing body.

The females are, in their way, more dangerous than the bulls because they are more unpredictable, quicker in their movements and changes of direction, and no less fierce. A half-ton bull that can accelerate from standing to 25 mph in a few seconds finds it hard to change direction once it has committed itself to charging at an object. That is why the safest bulls are those that, in Hemingway's words, run 'as if they were on rails'. The *ganaderos* (bull breeders) say, 'Bulls get their size and build from their fathers, but their hearts come from their mothers.' They gauge the likely ferocity and courage of their bulls by testing in an arena the two- to three-year-old females (*vaquillas*) that are going to be their mothers. This process is called the *tentadero*. It also maintains the rule that the bull must not see a man on foot before he fights. All the work with the bulls is done by mounted stockmen.

The *toro bravo* has two psychological states. He is relatively placid as part of a herd, but once separated and alone, the beast within is released. He tends not to show aggression unless threatened directly, but when upset, either with or without the herd, the instinct of both sexes is to attack. They will not back off. I saw a calf, less than 24 hours old, attack and see off a big dog. And I was told about a young woman whose thigh was broken when a calf attacked her. Alexander Fiske-Harrison describes his first encounter with a fighting bull, the pardoned Idilico, who was living out his days on the ranch of the Núñez del Cuvillo family between Seville and Cadiz. He describes what happened when the bull was put in a pen on his own. It 'was not a change of temperament, but of character itself. His head came up and a vast surge of testosterone-enhanced adrenaline seemed to course through him. He literally began to dance, as a boxer dances, his 1,212 lb bulk skittering on his hooves over the mud-slick stones.' When Fiske-Harrison climbed onto one of the six-foot-high walls surrounding the pen to look down, the bull caught sight of him and became an 'explosive paranoia of horn and muscle'. As he grasped the safety rail, Fiske-Harrison moved the little finger of his left hand and the bull flicked his head towards it, then his attention shot towards the little finger of his right hand when he moved that.

The bull does not react to pain in the normal way, by flight, but by attack. When wounded by the picador's lance, he redoubles his assault. In this, the type is unique among bovines. A bull will attack his fellows to gain supremacy in the herd or to defend his territory, and he never loses his predisposition to combat until the day he dies. A *vaquilla* behaves similarly when she is challenged. Fear turns into terrifying aggression. She will attack anything, even her own shadow, and will not stop until she has killed the thing that threatens her. This temperament has been bred into Spanish fighting cattle over many centuries, matching the bravest and most ferocious females with bulls that have shown extraordinary fearless aggression and indifference to pain in the arena.

Is it that these cattle have some special attribute of extreme aggression under pressure, something different from other breeds of cattle? Or would all cattle be like this if they had been refined and enhanced by domestic selection? Most domestic breeds have had the fight bred out of them because docility is needed to get the best from them. Whether or not Spanish cattle *are* descended from the aurochs (and no link has been proved), they have been selected for ferocity through countless generations (on average 20 every century) since Roman times – and probably a long way further back than that – so it is hardly surprising that the latest incarnation of these bovines should be such desperadoes. In fact, given that the bloodlines of the *toro bravo* are some of the oldest in Europe, it is more surprising that they are not completely unmanageable. They seem to have reached a plateau of ferocity and not to be any more dangerous now than they were 200 years ago.

Once a *toro bravo* has determined to charge, it will do so without deviation. That is why it can be fought with a cape. The Spanish call this tendency *la nobleza*: its nobility in not resorting to trickery such as stopping in mid charge or hooking or chopping with its horns; the bull is open, straightforward and honest. In other words, it plays by the rules and accepts its destiny as a player in the drama being enacted in the ring. This is, of course, pure anthropomorphism. The bull has no choice but to act in character. The rite of which he is part is a human construct that is supposed to mirror the ultimate and only reality: death in a tragic spectacle culminating in his ritual sacrifice.

There are many people (mainly in what could broadly be called the Protestant countries of the world) who deplore bullfighting and would readily see it banned. Ernest Hemingway thought they were that part of humanity that instinctively sides with the bull against the man and dislikes their fellow humans. I am not sure this analysis is right. You don't have to dislike mankind to have fellow feeling with animals and deplore wanton cruelty towards them. You can have empathy for both the bull and the man.

According to the Christian teaching upon which our European moral attitude to animals has been based for the last two millennia, we were given dominion over animals and the rest of the natural world. We are entitled to use them for our own purposes, including putting them to death for our food, even if it is simply to gain pleasure from eating them. But we degrade and brutalize our own nature and risk transferring the cruelty to humankind if we take pleasure directly in inflicting pain on animals. Animal suffering is repugnant to us to the extent that it reminds us of, or resembles, human suffering and therefore ordinary human kindness towards animals must be inspired by empathetic anthropomorphism.

Loving an animal cannot exclude killing it. If a dog is suffering, we would consider it a kindness to have it put down. That does not mean we do not love it; rather it means the opposite. A responsible owner will look after his dog and give it an opportunity to express its nature, as a companion or working dog, and then, in the proper exercise of owner's responsibility, will if necessary kill it at the end of its life. It is the same with farm animals. We keep cattle for a purpose: to give us milk, meat and other by-products. And it is the essence of their nature and existence that to fulfil their purpose, they have to die. That is not incompatible with loving them. And if we didn't keep them, there would be nothing to love anyway. A cattle breeder is not immune to sadness at having to send his animals to be slaughtered for meat. He is not a monster who gets pleasure from killing them.

When I was growing up, there was an old woman who lived alone in our village and kept a couple of pigs on the swill from the kitchens of a hotel she worked in. When killing day came round, she insisted on being present at the slaughter, as she had been for every pig she had ever reared. She never failed to weep over the death of a pig that had become almost a friend. But she didn't stop keeping them. She loved bacon and ham, black pudding and brawn and pig-foot pie

– and she loved her pigs, both for what they were and in gratitude for what their lives and their deaths gave her.

Rearing fighting bulls for killing in the arena seems to me no different from rearing a pig for bacon or cattle for beef. Those who do it almost certainly do not hate the animals they nurture so carefully, otherwise they would not make such a good job of looking after them. That their lives end in death says nothing about the feelings of the people who cause it. The great bullfighter Cayetano Ordóñez tells Fiske-Harrison about his sadness at killing, and the emotion a great bull can inspire in him: 'It is like a friend at that point [the point of death]. You do not want to kill it, but you have to, and that is your tragedy, your sadness. But it is *your* bull, only you can deliver death to it, for only you have risked your life to face it. And then, that, the moment of the kill, is the most important moment of all. For fifteen minutes the bull has been charging you, and now you must charge it with the sword. This is the only moment the matador himself charges the bull.' This is the *hora de verdad*, the moment of truth.

The *toro bravo* pays for his five years running free on the ranch with twenty minutes in the ring, just as the beef beast does with its eighteen months in the feedlot. 'The argument that killing for food is not the same as killing for entertainment is bogus,' says Fiske-Harrison. 'We eat meat because we like the taste to entertain our palates.'

The Miura *ganadería* is a haven for wildlife. There were flocks of birds, too far off across the fields to tell what they were, rising and wheeling in great clouds like swarms of bees and settling further away to work the soil for something they were eating. Across from the farm, a flock of partridge rose from the margin of one of the tarns that store water for the long dry season. I remarked on the wildlife and Don Eduardo ran through a list of species that their cattle rearing encourages, comparing this with the sparse fauna of the monoculture on the winter-brown arable lands we had travelled through.

The Miuras' wild pastures are part of the *dehesa* (*montado* in Portuguese). This covers about a quarter of the five to six million acres of the western and south-western parts of the Iberian peninsula. It describes an ancient integrated land use that gives the greatest productivity from arid marginal land in a Mediterranean climate. The soil is managed primarily for animals, both grazing and browsing: cattle, goats and sheep eat the natural mixture of native herbage. Black Iberian pigs feed on the acorns from the cork and holm oak trees that stud the land and go to make *jamón ibérico de bellota* (the famous dry-cured Spanish ham). They have a cultivating function, rooting up the ground and keeping it open and free from the scrub and oak saplings that would otherwise take over. The bark from the cork oaks is harvested every nine to twelve years, and the trees are spaced at just the right distance to balance the herbage's need for sunlight with maximizing the number of trees to make best use of groundwater and to yield the most acorns to feed the pigs and game. Although the herbage withers and dies in the heat of summer, it is resurrected from its dry roots by the rain that returns after the drought, and the land becomes carpeted once again with a billion wild flowers and viridescent foliage. The shade and roots of the oaks encourage fungi, bees make honey from the flora, and the branches shelter the Iberian lynx and the Spanish imperial eagle.

This ancient system involves the widest possible range of species, living in symbiosis, maintains soil fertility and gives considerable production from unpromising soils. Every species has a place in the tapestry; even the myriad flies that plague the cattle during the hot, dry summer sustain flocks of cattle egrets that patrol the herds as they lie cudding in the shade and peck the insects from their unconcerned hides. The manure from the cattle creates a living soil teeming with microbes; it sustains insects that feed ever-larger creatures, up to the dung beetles at the top of the insect pyramid, which do the heavy work of processing it. Each species lives according to its own cycle of life and interlocks with and supports the others. The oaks have the longest life cycle – about 250 years – and the tiny microbes the shortest. Every creature has a part to play in sustaining the whole.

This kind of extensive land use, which has evolved over thousands of years, provides a good enough living from poor soils to maintain and feed a surprisingly large population. The biggest cash income has traditionally been the sale of cork, but with the increasing use of plastic corks it is not as lucrative as it once was. High-quality dried hams also bring in a significant income. But it is those who rear fighting bulls who make the most from the *dehesa*. The best bulls will sell for over €20,000. One and a quarter million acres are used for rearing the 25,000 or so fighting bulls that are killed in the ring every year. As a significant proportion of the bull calves are not deemed courageous enough for *la corrida*, and half the calves are heifers, by far the greatest proportion of the offspring of the hundreds of thousands of cows kept to supply the bullfight are killed for their meat. This is a superior type of beef, lean and fed slowly on natural grazing. Not only are the wide lands of southern Spain maintained in their natural state by the income from *toros bravos*, but in the existence of the huge herds of semi-wild cattle that roam there, and in the bullfight, Spain's national pride and unique identity are preserved.

The bulls do not usually fight before they are four years old, and some are five, nearing six, when they enter the ring. Up to that age they live as naturally as it is possible for a domestic animal to live on extensive grazings. They are deeply attached to the place where they were born, where they are entirely unconfined as they grow towards their inevitable 20 minutes in the ring. Each bull is given a name and his life is dedicated by the matador to a saint, to the crowd in general or to someone important to the matador.

The origins of bullfighting go back into prehistoric bull worship and sacrifice. There are Palaeolithic bull paintings in numerous caves in western Europe and beyond. The first recorded bullfight is in the *Epic of Gilgamesh*, in which Gilgamesh and Enkidu fight and kill the Bull of Heaven: 'The Bull seemed indestructible, for hours they fought, until Gilgamesh dancing in front of the Bull, lured it with his tunic and bright weapons and Enkidu thrust his sword deep into the Bull's neck and killed it.' Like a matador, Gilgamesh drove his own sword into the bull's spine 'between nape and horns'. The oldest representation of what seems to be a man facing a bull is on a Celtic-Iberian tombstone from Clunia, near Burgos, in northern Spain. In Anatolia, excavations of a site dating from 6700 to 5650 BC have uncovered temples adorned with bull's heads and furniture and pillars made of stylized horns. Human-headed bulls were commonly carved into porticoes of important buildings in Sumer and Assyria. The bull god Apis was worshipped in ancient Egypt at Memphis, and Nandi the bull was revered in Hindu art and architecture. Theseus slew the Minotaur, and the central act in the cult of Mithras involves the slaying of a bull. The bull was a surrogate for the sacred king, sacrificed for the sake of his people.

The Spanish bullfight (where the matador almost always kills the bull) is a very ancient rite, maybe Thracian in origin. The Thracians were an early Indo-European tribe whose lands lay between the Black Sea in the east and the Aegean in the west. Greek and Roman accounts have them as especially warlike barbarians, ferocious and bloodthirsty. Plato said they were 'extravagant and high-spirited', but in this they were not much different from the Greeks' other warlike neighbours: Celts, Persians, Scythians, Iberians and Carthaginians.

There is some evidence that it was from the Thracians that the Emperor Claudius took bullfighting to Rome, and that he introduced it into Spain when he instituted a short-lived ban on gladiatorial combat. Robert Graves, in *The White Goddess*, suggests more plausibly that there may have been an earlier introduction into Spain, in the third millennium BC, by Iberian settlers who had cultural and racial connections with Thrace.

Perhaps the most intriguing depiction of interaction between humans and bulls is in the famous bull leaping fresco at Knossos. Sir Arthur Evans interpreted the activity as young people disporting themselves in a kind of ritual dance, vaulting and leaping over a bull. But there is another interpretation put forward by a German geologist and author, Hans Georg Wunderlich, who examined the structure of the palace at Knossos and published his findings in English in 1975 in *The Secret of Crete*. He concluded that the building could never have been a palace for the living and was more likely a charnel house, a huge necropolis, containing an arena where human sacrifices took place. The 'bull vaulting game' was in fact a ferocious form of human sacrifice, which involved young men and women being gored and trampled to death as a ceremonial offering to a sacred bull, and was the origin of the legend of the Minotaur. Since then, Wunderlich's findings have found support from other scholars.

Although bullfighting is often linked back to Rome, where human contests against animals were commonly held for entertainment, it has deeper pagan roots. It is not a coincidence that some of the oldest bullrings in Spain are built on or near the sites of Mithraic temples. The early Church's foremost rival in faith was the mysterious and powerful cult of Mithras, the pagan god of Persian mythology, widely worshipped in ancient Rome, especially by soldiers. The killing of the sacred bull (tauroctony) is the central symbolic act of the cult, and was commemorated in the Mithraeum wherever the Roman army was stationed. The ceremony of bull killing was depicted in art and stone throughout the Roman Empire, even as far north as Hadrian's Wall, where a cave-like temple to the cult of Mithras was found in 1949 at Brocolitia, near the Roman fort and important crossing point on the North Tyne at Chesters. The early Church was not sympathetic to the bull, likening it to the devil. At the Council of Toledo in 447, the devil was described as 'a large black monstrous apparition with horns on his head, cloven hoofs, hair, ass's ears, claws, fiery eyes, gnashing teeth, a huge phallus and sulphurous smell'. This could have described any of the bulls depicted in representations of the Mithraic sacrifice.

As early as the third century BC, Iberian cattle were known to be different from other domestic cattle. They were ferocious, with an instinct to attack without provocation, and would try to kill their adversary. The Iberians used these wild cattle in warfare. In 228 BC, the defenders of a town besieged by Hannibal's father, Hamilcar Barca, gathered a herd of the wild cattle, harnessed them to wagons loaded with resinous wood lit with torches and drove the ensemble at the enemy. Barca was killed and his army destroyed. Even at this date in Baetica (Andalusia), games were being held in which men showed bravery and skill with bulls before killing them with a lance or axe.

This continued after the Visigoths conquered Spain in AD 415, with spectacles and games involving men pitting their strength against fierce bulls. After the Muslim invasion and annexation of Andalusia in 711, these evolved into mounted bullfighting contests between Moorish chieftains and Christian knights, either in arenas, the city square or open fields outside the towns. By the end of the eleventh century, these fiestas of bullfighting were well established, particularly in the south of Spain, and have continued into present times. The running of the bulls (*encierro*) during the Fiesta de San Fermín in Pamplona is probably the best known. Numerous people are trampled and gored, and yet its popularity only seems to increase. People (mostly men) come from all over the world to dress up in white shirt and trousers, tie a red sash around their waist and take their chance against a herd of bulls stampeding through the narrow streets towards the arena where they are going to be fought in the afternoon.

Religious festivals and royal weddings were celebrated by fights in the local plaza, where noblemen would compete for royal favour while the common people enjoyed the spectacle. Until the eighteenth century, bullfighting in Spain was reserved for the nobility. A single mounted knight armed with a lance fought a bull in a closed arena. This goes back at least to the time of Charlemagne (ninth century). The first Castilian to lance a bull from horseback is supposed to have been El Cid (1043–99). There is a record in a chronicle from 1128 that there were bullfights at Saldañato to celebrate the marriage of Alfonso VII of León and Castile to Berengaria, daughter of Ramon Berenguer III, Count of Barcelona. By the time of the Austrian accession in 1516, bullfighting had become a necessary part of every court occasion. Charles V endeared himself to his subjects by celebrating his son Philip II's birthday with the lancing of a bull.

For over 600 years the fight was the preserve of mounted noblemen with a lance, later a short spear, fighting a bull that had been manoeuvred into position in the ring by men on foot luring it with capes. The House of Bourbon disapproved of bullfighting, so when Philip V succeeded to the Spanish throne in 1700, the aristocracy gradually abandoned their part in the mounted spectacle. The Crown's disapproval had little effect on the wider public enthusiasm, and in a reversal of roles, the mounted matador became the supporting player (the picador), and the man on foot took the leading role as the matador. In Portugal, the mounted spectacle continues, with the bull being lanced from horseback and then wrestled into submission by a team of 'bull-grabbers'.

There is a spiritual aspect to the ceremony, which fuses ancient pagan superstition, myth and ritual with Christianity. It is possible to see in it a symbol of the tension between nature and nurture, unredeemed barbarism and Christian revelation, concern for God's creation and human salvation. But the medieval Church did not see it as anything of the kind.

In 1567, Pope Pius V banned bullfighting, excommunicating Christian noblemen who participated in or facilitated it and refusing Christian burial to anyone killed in the ring. This had little effect on popular Spanish passion for the *corrida*, and eventually the Church was forced to relent. Bullfights became part of the fabric of Spanish social and community life, inserted into the Christian calendar and held on feast days and saints' days; indeed, in many places, the opening day of the bullfighting season is Easter Sunday. And every bullring has a chapel, where the matador goes to pray and can receive the sacraments (including extreme unction) before he goes into the ring.

Spain's oldest stone bullring, at Ronda, was built in 1785 in a surprisingly intimate neoclassical style on a spectacular site almost on the cliff edge on which the town perches. The sandy sienna floor rises gently from the sides into the centre, and standing there you realize how terrifyingly far away you are from any refuge. Even when the stalls are empty and there are no bulls, it is an eerie and unnerving experience and gives an idea of the raw courage needed to face an animal determined to kill you. The size of the arena floor is the same in every bullring, whatever the number of seats for spectators; the only exceptions are those at high altitudes, which are slightly smaller to compensate for altitude fatigue.

Hemingway said the bullfight is the only art form in which the artist risks death every time he practises it. He might be on to something here about the Spanish fascination with death and its nearness. The Caballeros, my daughter-in-law's family, own a hacienda near Seville where they grow oranges, almonds and olives. It is part of a tract of land given to one of their ancestors by the Spanish king, along with the title of *marqués*, for trouncing the Portuguese in some conflict or other a few hundred years ago. The hacienda is built in the Andalusian style in a square courtyard just like the Miuras'. At one time they kept a large herd of cattle. The main difference between this place and a similarly sized English estate (1,200 acres) is that the family's ancestors are still here, not in a nearby village church, but interred in the walls of the courtyard. They lie in the mortuary of a little chapel in what look like pizza ovens, a constant reminder of the inevitability and closeness of death.

CHAPTER 19

Sacred Cows

'O Mary, go and call the cattle home,
And call the cattle home,
And call the cattle home,
Across the sands of Dee.'
The western wind was wild and dank with foam,
And all alone went she.

Charles Kingsley, 'The Sands of Dee'

THERE IS A pervasive modern Western view that if only humans would leave the natural world alone, it would balance itself out in perfect harmony. In fact, it would be better if humans did not exist at all, but seeing as they do, their numbers should be limited and their activities confined to as little of the natural world as possible. This is the main justification for the so-called rewilding movement. The people who support it are usually vociferous opponents of bullfighting. They cite cruelty, barbarism, causing suffering to an animal in an unnecessary rite and failing to respect its dignity – which is more or less what was done to the animals in a Dutch experiment of 'rewilding'.

The Oostvaardersplassen is a unique Dutch 'nature reserve', 25 miles east of Amsterdam, on 22 square miles (about 15,000 acres) of almost completely flat, fenced-in polder, reclaimed in the 1950s and 60s, much of it well below sea level and only kept relatively dry by sophisticated drainage and pumping systems. It forms part of the recently created twelfth Dutch province of Flevoland, more or less in the centre of the country. Since the end of the last Ice Age, the area had lain at the bottom of the Zuiderzee until a huge network of dykes was dug in the 1930s to keep out the sea and transform it into a freshwater lake. Later it was drained to create Flevoland from the rich silt of the lake bed.

Despite being some of the most fertile soil in Europe, where almost anything would grow, while the soil was drying out the Dutch government was persuaded by a group of conservationists and ecologists to turn part of the reclaimed province into a 'Paleolithic' landscape in a grand experiment to return it to the state in which they believed it would have been at some point in pre-history – had it not, of course, been under water. They introduced creatures they believed would have naturally populated the northern European landmass before mankind dispossessed them. And they left them alone in their enclosure to see what would happen. There was no management or feeding; the animals were to take care of themselves in a 'natural' existence.

Where some of the species had become extinct they had to settle for the next best thing. And in the case of cattle, as they had no aurochs, they brought in some Heck cattle. These multiplied, as did the red deer, which had been cap-tured in Scotland, and the horses, imported from Poland, and the foxes and the wolves. In fact, all the large mammals reproduced so prolifically that they formed what could, with a certain amount of rose-tinted licence, be said to resemble the great migratory herds of the African plains. The German magazine *Der Spiegel* called the Oostvaardersplassen 'the Serengeti behind the dykes'. Visitors paid up to €40 each to take tours of the reserve.

As the animals proliferated, they began to starve. The founders refused to do anything about it, insisting that nature must be allowed to run its course, because nature knows best. Many animals collapsed from exhaustion, or drowned in the wetlands, too weak to clamber out. The birds survived because they could fly away when the food ran out and seek sustenance elsewhere. During the harsh winter of 2005, commuters travelling into Amsterdam were greeted by the spectacle of emaciated cattle, deer and horses crowding against the perimeter fence, desperate for food, while foxes and corvids harried and preyed on them. They died in droves like inmates in an extermination camp, unable to escape, their carcasses picked clean by the predators with which they were forced to share their miserable captivity.

This caused a public outcry and forced the State Forestry Service, which is nominally in charge, to introduce the culling of animals that the wardens judge to be too weak to survive the winter. Between 30 and 50 per cent of the large herbivores are now shot before they can starve to death. But all the land that can be grazed is still eaten back to the sod, with only a few unpalatable species, like dandelions, showing above the ground. And to try to satisfy their hunger, the wild horses and deer have killed hundreds of thousands of trees by stripping the bark off them. The trees' rotting remains are spread over huge areas, which look as if a hurricane has swept over and flattened them. The loss of shelter that the trees gave the hungry grazing animals has resulted in even more cruelty during the winter, when they have nothing to eat anyway and are trying to survive off their accumulated fat. There is no chance of regeneration of the trees, because every seedling is eaten almost before it pokes through the soil. This is an ideologue's paradise. Any criticism of the mess and cruelty that is being perpetrated is condemned as 'middle-class aestheticism'. Why should a landscape have to conform to any particular canon of beauty? Leaving land 'natural' is its ideal state and any human intervention illegitimate.

'Rewilding' is meaningless, based on ideology rather than reality or historical accuracy. The assumption is that at some time in the past, northern Europe (and by implication the rest of the world) supported a range of wild animals that lived in a kind of ecological harmony and that it would be a good thing if such an arcadia could be recreated. But nobody knows how things were in the Paleolithic era, and in any event, many of the species that might have existed then no longer do. 'Wilding' would be a more accurate way to describe it. But even then, its supporters ignore the fact that the whole 'reserve' is just that, a reserve, fenced round and kept dry artificially by electric or diesel-powered pumps.

The ecologists and conservationists seem blind to the truth that nature left unmanaged produces anarchy, deserts, lingering cruel deaths and misery. They seem to believe that nature is benevolent and wise, and don't recognize or even care that allowing it to run unchecked inevitably leads to horrific results. Nothing as extreme as Oostvaardersplassen is being proposed (yet) by ecologists in Britain, but that is perhaps because a radical application of wilding would tend to put people off. It is better to approach such things by degrees so that people don't notice. Although in typically provocative fashion, environmental activist George Monbiot has praised the Oostvaardersplassen experiment and published a manifesto to 'rewild the world'. Of course, Monbiot and his urban ideologues will not be the ones to suffer when all the farmers and country people have been cleansed from the land. One farmer I was talking to about Oostvaardersplassen pointed out that if he kept cattle in these conditions in winter and allowed them to starve to death with no shelter, he would be prosecuted and probably banned from keeping animals again. Why should these rewilding ideologues be treated differently?

This Dutch experiment inspired a new movement – Rewilding Europe – with the belief that what they call 'new nature' can be created. As a result, every year for the last few decades, encouraged and paid for by the EU, tens of thousands of acres of marginal farmland in Europe have been abandoned to 'nature' to achieve something that generations of people in the past would have thought ridiculous: replacing wilderness that we have apparently lost. The same thing is proposed for the depopulating expanses of the American Midwest. And it looks as if the British government intends to try something similar with tracts of British countryside that are deemed not to be needed to produce food or accommodate the people who live there. This seems to me a denial of our sacred trust to care for creation, to manage the natural world by culling over-populous predatory species and encouraging the weaker endangered ones.

The Oostvaardersplassen reserve is by no means the first effort to enclose cattle and leave them to it, although it was the first where the enclosers watched the cattle starve to death. Land has long been enclosed for game reserves and pleasure parks by landowners who can afford to keep animals or just arrange the landscape to suit their fancy. Enclosing land around a gentleman's seat to give him privacy, or a place for his family to disport themselves, or to contain a herd of deer or other game, became very popular in England after the Crown relaxed the forest laws. At one time there were nearly 2,000 such private parks, most of which have not survived changing fashions or financial stringency.

In 1225, Henry III enacted the Charta Forestae, which removed many of the restrictions on land imposed by William the Conqueror. Over the century and a half following the Conquest, about a third of the land had become concentrated under Crown control. The Crown had the dominant right to use the land to preserve wildlife and for hunting. The prospect of penalties such as death, blinding and mutilation discouraged people from breaking the forest law – mainly taking game – though it was hard to catch miscreants. As William I had taken England by conquest, in feudal law it became his absolute possession and he was entitled to dispose of it as he saw fit. He declared large areas to be 'forest', which meant that the occupants of the villages and towns that lay within these areas found themselves subject to a set of rules that overrode the rights they had enjoyed under the common law. This limited their freedom to use the land they occupied for their own benefit and incidentally discouraged improvement. A present-day analogy, although it cannot be taken too far, are the rules in the Wildlife and Countryside Act 1981, which curtail many of the rights that occupiers of land had previously enjoyed, for the apparent purpose of preserving wild flora and fauna. The statute overrides any common-law rules that conflict with it and that previously obtained.

One of the effects (probably unintended) of releasing land from direct Crown control was to free certain larger land-owners to enclose private parks, some of which they stocked with cattle for the sport of hunting the 'wild' bulls. Herds of cattle were established in many places, particularly at Blair Atholl, Cumbernauld, Drumlanrig and Cadzow in Scotland, and Barnard Castle, Chartley, Hoghton, Chillingham and Lyme Park in England. Even Windsor Forest had a herd of white cattle established in 1277.

Four hundred years ago, the herd at Hoghton had the honour of giving rise to the naming of a cut of beef that has become part of the language. In early August 1617, King James I found himself travelling south through Lancashire. He had a habit of descending upon his richer subjects and expecting liberal hospitality for his entourage. He lit upon Sir Richard Hoghton at Hoghton Tower, near Preston, who slaughtered for the occasion a beast from the herd of White Park cattle enclosed in his park. At the banquet James declared: 'Finer beef nae man ever put his teeth into. What joint do ye ca' it, Sir Richard?'

'Sire, it is a loin of beef,' replied his host.

'A loin! By my faith, that is not a title honourable enough for a joint sae worthy. It wants a dignity, and it shall hae it. Henceforth it shall be Sir-Loin, an' see ye ca' it sae.'

And of course we have ever after obeyed the king's injunction.

It is almost certain that the cattle enclosed in these parks were the same kind of white cattle that had been common across the west and north of England, Ireland and Wales for many centuries and are Britain's oldest native breed. The 330-acre park round Chillingham Castle in Northumberland is one of the few of these gentleman's parks that has survived. Its herd of white cattle with red ears are examples of the ancient British white cattle, with the added distinction of being, to a large extent, semi-wild because they have been mostly left to their own devices for a long time, though there has always been a steward or warden to keep an eye on them and feed them hay in winter if the snow is too deep to forage. The herd was reduced to eight cows and five bulls during the awful winter of 1947, and until 2001 its size was limited by having to share the grazing with a flock of sheep. When these were removed, the herd increased rapidly from about 40 to over 100 animals. It remains to be seen whether the grazing will be adequate to sustain such a large herd.

According to records kept between 1862 and 1899 on the suggestion of Charles Darwin, it was the practice to castrate 40 per cent of the bull calves and shoot about 12 per cent of the cows for meat – presumably the poorer bulls and the barren cows – and occasionally one would be put out of its misery if injured or ill. Until recent times they might also hunt a bull or two, if the herd was getting too big for its grazing and some culling was deemed necessary. Word would be passed around locally and people would turn up on the appointed day, on horseback or foot, in a state of great excitement. The animal to be culled would be ridden off from the rest of the herd and then shot by a marksman. In 1872, Edward, Prince of Wales, was staying at Chillingham, recuperating from typhus and, for 'a bit of sport', was allowed to shoot the dominant 'king bull'. Concealed in a hay cart, he killed it from a distance of 70 yards with one shot. The head was then mounted for display at Sandringham. It somehow made its way into the lounge bar of The Bull (also known as The Chillingham Arms) in Jarrow, but has not been seen since the pub was demolished in 2005.

The Grey and Bennet families have owned Chillingham since the late thirteenth century, and have been proud and protective custodians of their cattle, which is one reason why, almost alone as a herd of white cattle, they have survived into modern times. Another reason is the remoteness of their domain. Over the last two or three centuries the herd has caught the imagination of many influential people, who have seen in it a reflection of their own particular preoccupations. Darwin used it to support his theories of natural selection, believing that the cattle were the direct descendants of the aurochs, the '*Gigantis primogenis* race ... which was described by Caesar in the forests of Germany'. Others responded to the romantic call of the noble savagery of ancient wild cattle apparently untouched by human influence since beyond the memory of history. An early enthusiast was Marmaduke Tunstall from Yorkshire, whose own family had lost a herd of similar white cattle to rinderpest. He explained to Joseph Banks, president of the Royal Society, in 1790 that the herd were 'Ancient Britons', aboriginal, dignified and 'noble', 'once the unlimited rangers of the great Caledonian and British forests' and retained 'much of their native fierceness'.

Thomas Bewick was inspired by a commission from Tunstall in 1789 to produce an engraving on wood, *The Wild Bull*, which he is said to have regarded as his masterpiece. Bewick became friendly with John Bailey, the Chillingham steward, and came several times to sketch the cattle. In their bucolic grandeur, apparent freedom and indomitability, he saw a symbol of a pastoral England fast being destroyed by industrialization and the flight to the towns.

Sir Edwin Landseer's magnificent painting, *Wild Cattle of Chillingham*, did much to bring the herd to public attention in the 1830s and touched the romantic imagination of an early Victorian Britain that saw the primeval freedom and purity of the cattle as a counterpoint to their own increasingly prosaic and utilitarian lives. The painting also represents the ideal family, a combination of the masculine readiness to fight to protect the nurturing mother and both parents caring for their vulnerable calf, which is symbolically placed in the foreground of the painting.

Wild white cattle stimulated the romantic interest of Sir Walter Scott, who wrote about them in his poem 'Cadyow Castle', published in 1802.

Mightiest of all the beasts of chase
That roam in woody Caledon
Crashing the forest of his race,
The mountain bull comes thundering on
Fierce, on the hunter's quiver'd band,
He rolls his eyes of swarthy glow,
Spurns, with black hoof and horn, the sand,
And tosses high his mane of snow.

He is referring to a similar herd of white cattle in Cadzow Park belonging to the Duke of Hamilton and also established in the fourteenth century, about the same time as Chillingham. The herd was moved to the duke's seat in East Lothian in the 1960s. Scott also had the Master of Ravenswood save Lucy from a wild white bull in *The Bride of Lammermoor*, where he does mention the white cattle of Chillingham, 'descendants of the savage herds which anciently roamed free in the Caledonian forests ...'

But as good as the tale is, and as much as eighteenth- and nineteenth-century romantics would have liked to believe that these white cattle were the direct descendants of the aurochs, it is simply not the case. Every breed of domestic cattle is of the same species, *Bos taurus*. Modern DNA analysis has been unable to make any connection between genetic material from the remains of what are believed to have been aurochs found in Britain, and the Chillingham cattle (or any other British breed of cattle, white or otherwise). But this does not dampen people's fascination with these beasts, or diminish their admiration for their supposed ancient origins.

In one sense, white cattle *are* an ancient type, not as the offspring of wild animals, but as the descendants of domestic cattle that have been in Britain for a very long time – probably longer than any other. We do not know how long or where they came from, but this has not prevented various theories (fanciful or otherwise) from being promoted about their provenance. One common one is that they came with the Romans; another that they were kept by the druids for ritual sacrifice, and when Christianity supplanted druidism, they were released into the wildwood and became feral. There may be some truth in the second theory, but it is likely that white cattle very much pre-date the Roman invasion.

Pliny writes about a druidic sacrifice of two white bulls in AD 43, fully a decade before Caesar's legions invaded. Numerous other literary sources refer to white cattle with coloured ears (black or red) at least two millennia ago throughout northern and western Britain, particularly

Ireland and Wales. The cattle mentioned in the Irish epic, *The Cattle Raid of Cooley* (*Táin Bó Cúailnge*) are white with red ears. There are also accounts of white cattle in Kilkenny in the early eighteenth century. This could mean either that they were ubiquitous, or they merited mention because they had a special or mystical significance and were reserved for ceremony or ritual. There are also clues in certain place names: for example, Inishbofin, off the coast of Connemara, means 'the island of the white cow'. Of course, naming an island after a white cow could mean that white cattle were unusual, rather than common.

There are similar references to white cattle in Wales. Rhodri Mawr, a Welsh king, who built Dynevor Castle as a defence against the Vikings, is recorded as keeping white cattle in AD 856. His grandson, Hywel Dda, formulated laws under which fines for certain offences were paid in cattle to the lords of Dynevor Castle, the administrative and military capital of Wales from about AD 800. Nine generations later, Rhys ap Gruffydd, who ruled the kingdom of Deheubarth in South Wales, and who died in 1197, was keeping colour-pointed white cattle. In the Holinshed Chronicle entry for 1211, the wife of the Norman baron William de Breos (Brecon), in an unsuccessful effort to appease King John, sent a gift of 'four hundred kine and one bull, of colour all white, the ears excepted, which were red'.

The nearest modern relatives of these ancient white British cattle (represented by the Chillingham and Cadzow herds) are the horned White Park, the polled British White and another closely similar horned type, the Vaynol, of which there are only three herds in Britain. They all have the same distinctive points (although black, not red) on the ears, nose, eye rims, feet, teats and tips of their long horns. The Chartley herd of White Park cattle, now kept at Ditchingham, in Norfolk, has been in the Shirley family since 1248 (although it was dispersed in 1905 and then re-formed in 1970). The head of a Chartley White Park bull was adopted as the emblem of the Rare Breeds Survival Trust. All the White Park cattle are of the same lineage as the Chillingham cattle, but the latter have been kept as a closed herd, isolated and semi-feral, whereas the White Park herds have all exchanged stock and introduced blood from other breeds at various times and are more or less docile. As a result, the Chillingham cattle are recognized as a separate breed in their own right, which is technically correct, because they are the only one of the three to have red points, particularly the ears.

As with many of the older breeds, the White Park are exceptionally thrifty and indiscriminate grazers of all kinds of vegetation, which makes them ideal 'conservation' grazers. This is a new way to describe old knowledge that any farmer would have imbibed almost with his mother's milk. Grazing a sward with old breeds of cattle improves its quality and productivity and makes its stems tiller out to cover the soil, because the cattle do not favour one type of grass or plant over another and therefore none of the sward grows to seed and, in theory at least, it all remains young.

One great disadvantage of enclosing a herd is the effect of continual inbreeding. The Chillingham cattle have been isolated in their park for longer than any other British park cattle. Classical genetics would tell us that harmful recessive genes would manifest themselves quite quickly and after only a few generations would kill off the population. Why this has not happened with these cattle is unclear. One theory is that

5 per cent of animals in a closed herd will survive inbreeding and the recessive genes will have been 'purged' from the population. This has probably happened at Chillingham, although they have retained (along with other White Park cattle) a relatively harmless congenital dental defect that causes some of their molars to be misaligned. There will also have been a loss of some beneficial characteristics: for example, the cattle are smaller than their White Park relatives, which has probably occurred because it helps them to survive hard times; and there could have been a loss of bull fertility, but that is just speculation.

One thing that concerns their keepers is that although the herd is tough enough to survive harsh weather and starvation rations, they may not have resistance to infectious diseases because of their isolation. That, and fear of having to slaughter the herd if foot and mouth broke out nearby, persuaded the trustees who now own the cattle and the park to establish a reserve herd in the north-east of Scotland in 1970. And during the Second World War, the threat of invasion persuaded Winston Churchill to send some of the herd to Canada for safe keeping.

It seems from the available sources that in Celtic folklore and culture white cattle with red ears were special, 'fairy cattle', owned and controlled by the inhabitants of the other, unseen world. One of the most, if not *the* most important of the early Irish epics is *Táin Bó Cúailnge*. In this, two great bulls, Donn Cúailnge (the Brown or Dun Bull of Ulster) and Finnbennach (the White-horned Bull of Connaught) represent the fortunes of their respective provinces. Their conflict runs through the narrative, climaxing with the vanquishing of Finnbennach by Donn Cúailnge. They are of supernatural, probably divine origin, as are the bulls that appear in Celtic oral tradition in Ireland, Scottish, Manx and Galician folklore.

White cattle with red ears belonged to the Sidhe, the elemental powers from an alternative dimension – fairies, as they're commonly described – who kept herds of them, usually under a lake or beneath the sea, where they grazed on seaweed. Occasionally one of these fairy cattle would appear on the shore as a gift for a human farmer, and if he treated it with respect and gratitude it would give him boundless milk. Scottish legends have these fairy cattle left as gifts, and they can be identified by their red ears. In Ireland, a sacred white heifer would appear somewhere every May Day to bring luck to a certain farmer. On occasion, the fairies would insert an enchanted cow into a mortal herd, and if the farmer was not vigilant in preventing his cattle from following it, it would lure them into the Otherworld through a fairy mound or knoll and they would never be seen again. It is said that harming a white cow with red ears will result in the Sidhe avenging their beast with the death of the malefactor. According to another old belief, if a man should give a cow to the poor, at his death the spirit of the animal will come to guide him to Paradise. White cattle were a particularly powerful symbol of purity, and their milk even more so. It was credited with healing properties for wounds suffered in battle and as an antidote to poison. Its efficacy was enhanced because the cattle came from the Sidhe.

A Donegal story tells of a family with ten children who were starving. One dreadful stormy evening there came a lowing outside the door, and standing there was a white cow with red ears. The father let her into his byre, and next day she gave birth to a fine heifer calf and started to produce copious amounts of milk. They asked around to find if anybody had lost a cow, and when nobody claimed her, the family kept her. Over the years, she gave birth to more calves and continued to give abundant milk that fed the family. One day the father found her grazing and trampling a field of oats, and forgetting the good fortune she had brought, chased her off with his stick. She gathered up her little herd and disappeared and was never seen again.

There is an ancient and powerful human taboo against eating carrion. But only in Ireland could it be turned into a supernatural story concerning the Sidhe. When a cow died suddenly, the owner would not touch its meat, or any part of the beast, but would organize a burial party to inter it as quickly as possible, crossing themselves as they walked away from the hastily dug grave. Its sudden death could only be explained by the real cow having been stolen away by the Sidhe into the next world, leaving behind the appearance of a cow that had pined away and died because its substance had been stolen. The real cow had been transmuted by a glamour, a spell, and the owner saw only a dead cow rather than what was really there: a horrible corpse from the Otherworld that was almost too dangerous to touch, never mind eat.

Another sort of carrion that must never be touched or eaten is calamity meat. This is the flesh of a beast that has died unexpectedly from some unexplained cause – maybe an accident. In reality the animal has been stolen by the Sidhe and taken to add to their herds in the Otherworld. The remaining carcase is not actually a cow, but a piece of alder wood magically fashioned to look like one. This must never be eaten because it is enchanted and will do harm.

Many of these cautionary tales have survived into modern times in Ireland. They run too deep in the culture to be eradicated by a few decades of scientific rationalism. One modern story concerns a young woman who kept a few sheep for their wool, which she spun into yarn. When one died unexpectedly (as sheep are wont to do), she told her old wise-woman neighbour that she was going to shear the dead sheep before she buried it. The neighbour was horrified and warned her against it in strong terms. She must bury the carcase, fleece and all: nobody would ever wear a garment made from such wool because it would be enchanted and harm the wearer.

We might be surprised to learn that the most valuable thing we can get from cattle is neither milk, nor meat, nor motive power, but the calcified bolus of hair from one of their stomachs. Bezoar stones, sometimes found in the reticulum (second stomach) of a ruminant, can still be worth ten times their weight in gold. The common-law rule of *caveat emptor* ('let the buyer beware') that had endured for many centuries was confirmed in a case concerning a bezoar. In 1603, one Lopus paid £100 to Chandelor, a goldsmith, for what Chandelor claimed was a bezoar. It turned out not to be a bezoar, and Lopus sued for the return of his money. He won in the lower court but lost on appeal because although Chandelor knew the object was not a bezoar, he didn't guarantee it to be one; he merely asserted that it was: 'The bare affirmation that it was a bezar-stone, without warranting it to be so, is no cause of action: and although [the defendant] knew it to be no bezar-stone, it is not material; for everyone in selling his wares will affirm that his wares are good, or the horse which he sells is sound; yet if he does not warrant them to be so, it is no cause of action, and the warranty ought to be made at the same time of the sale.' Lopus lost his £100 (more than £30,000 in today's money) because he ought to have looked out for himself.

But why did Lopus pay ten times its weight in gold for what he believed to be a calcified bolus from the reticulum of a ruminant? The commonest source of bezoars was the intestines of wild goats from Persia; hence their name, derived from the Persian *padzhar* – *pad*, meaning 'protector', and *zhar*, meaning 'poison', thus 'counter-poison' or antidote. It had been believed from ancient times that a bezoar was a universal antidote to any poison and simply immersing one in a drinking glass would neutralize poison drunk from it. The Andalusian Muslim physician Avenzoar (Ibn Zuhr, d. 1161) records the widespread medical use of bezoars as a panacea in the twelfth-century Arabic book of magic and astrology *Picatrix*, which contains extensive references to their efficacy. The Crusaders brought them to Europe from the Middle East in the eleventh century.

Medieval monarchs commonly used bezoars. The emperor Rudolf II of Prague wore numerous stones as an amulet against his chronic melancholia. He had a bezoar made into a drinking cup mounted in enamel and gold and studded with rubies and emeralds. They were valued as a purgative, and even thought to protect against the plague. Napoleon Bonaparte was sent bezoars during his confinement on St Helena. Had he not thrown them into the fire, he might not have succumbed to the arsenic that is said killed him. Elizabeth I took their protective powers seriously. She wore rings inset with bezoars and immersed one in her cup before she drank. Some rulers kept a stone set in a gold filigree case on a chain around their neck, to be ever available to suspend in their wine.

The stones remained popular until the empirical spirit of the Enlightenment caused people to question their efficacy. The French surgeon Ambroise Paré famously doubted their protective properties, and in 1575 took the opportunity to carry out a gruesome experiment. A cook at the king's court was caught stealing silver and sentenced to hang. Paré offered him the alternative of taking poison with the protection of a bezoar, and if he survived, his life would be spared. The poor man died in agony seven hours later.

Although news of the unfortunate experiment somewhat dented their popularity, that was not the last word on bezoars' curative properties. Modern research has shown that when immersed in an arsenical solution, a bezoar will absorb the toxic compounds arsenate and arsenite. So the Arabs were half right. Bezoars are effective against arsenic, which was the poison most likely to be used against a medieval monarch, but not as the cure-all claimed for them.

Bezoars are still highly prized in Eastern, particularly Chinese, medicine and as curiosities in the West. One recently changed hands in London for £29,000, although it was beautifully set in gold with a gold chain. And of course, Harry Potter saved Ron Weasley's life by pushing a bezoar down his throat when he was poisoned by drinking oak-matured mead intended for Professor Dumbledore.

If we are ever tempted to overlook the central place occupied by cattle in human society, we only have to bring to mind the billion or so people in India who claim the cow to be sacred. Hindus believe that all creatures have a soul, but particularly the cow, which God created as the first creature immediately after mankind. Therefore killing one would be no less sinful than killing a human being. In the Sanskrit epic the *Mahabharata*, written between 300 BC and AD 300, the transition from hunter-gathering to settled farming is mythologized: 'Once when there was a great famine King Prithu took up his bow and arrow and pursued the earth to force her to yield nourishment for his people. The earth took the form of a cow and begged him to spare her life; she then allowed him to milk her for all that the people needed.' Gandhi used this image of the earth cow for the ideal of an Indian nation sustaining her people.

The Hindu taboo against eating beef is tied up with social status. The higher the caste, the greater the food restrictions and the more starkly its members are differentiated from other castes and faiths. Rather reminiscent of the Western concern with 'clean eating' and the recent upsurge in veganism, it is tied up with female purity, docility and generosity – the cow gives her milk from her own body without stint – and the milk's purity and healthfulness give those who consume it a higher moral value than those who are less discriminating in their diet.

In the last few years, the cow's sacredness to Hindus has also become a rallying point around which growing nationalism and hostility to Muslims and other minority religions is coalescing. Hundreds, even thousands of 'cow vigilante groups' have formed all over India, nominally to protect cows from being harmed, but actually as a powerful symbol of popular resistance, given form in Hindu nationalism, to increasing state domination of Indian custom and society, and what is seen as creeping Islamization. They are also claiming to enforce the cow protection laws that are in force in almost all of India's states. Breach of these is punished with some severity. For example, in Gujarat, life imprisonment and a large fine is imposed for killing a cow. And in 2015, again in Gujarat, lower-caste Hindus were punished by flogging for skinning a dead cow. This started street protests and led to the resignation of the state's chief minister.

The groups, mostly young men, are often armed and act as vigilantes, attacking Muslims whom they suspect of harming cattle, smuggling them for slaughter or otherwise dealing with their carcases. This is part of the Hindutva movement ('Hinduness'), which started in 1923 and has widespread support among the Hindu population and the tacit encouragement of a large section of the Bharatiya Janata Party (BJP). Reuters reported in June 2017 that 28 people had been killed and 124 injured in 63 attacks since 2010 in 'cow-related violence'. The scale of the attacks is described as 'unprecedented'. In June 2017, a popular Hindu preacher, Sadhvi Saraswati, said that anyone who ate beef should be publicly hanged. One Hindu animal rights activist claimed that the slaughter of cows was directly responsible for global warming because it released 'EPW' – emotional pain waves. After September 2014, the government banned the slaughter of buffaloes, in a move that struck a blow against the Muslim-run beef industry.

It is a particularly Indian response to religious differences for the modern nation to coalesce around their ancient reverence for the cow, a potent religious and nationalistic symbol capable of arousing deep emotion. It is also a powerful acknowledgement of the central place that cattle still have in human life in India, which would be unthinkable without the cow.

But it would be similarly unthinkable in Europe or America, or anywhere else in the world, whether we in the urban West recognize it or not. For cows are not just utilitarian providers of flesh and milk; they are creatures that are enfolded in our very existence. There is hardly an aspect of human life that is not touched by our association with cattle, or enriched by the gifts they give us. Until the last century, throughout all recorded time, the patient cow is the one animal we could not have done without. It has been with us from the beginning; it was there at the birth of Christ; it is entwined in our myths and folklore. For wherever we have journeyed through our long history, always at our side has been the long-suffering ox, plodding doggedly on.

The black one, last as usual, swings her head
And coils a black tongue round a grass-tuft. I
Watch her soft weight come down, her split feet spread.

In front, the others swing and slouch; they roll
Their great Greek eyes and breathe out milky gusts
From muzzles black and shiny as wet coal.

The collie trots, bored, at my heels, then plops
Into the ditch. The sea makes a tired sound
That's always stopping though it never stops.

A haycart squats prick-eared against the sky.
Hay breath and milk breath. Far out in the West
The wrecked sun founders though its colours fly.

The collie's bored. There's nothing to control ...
The black cow is two native carriers
Bringing its belly home, slung from a pole.

Norman MacCaig, 'Fetching Cows'

Glossary

acre: 4,840 square yards and the amount of land a man and an ox could plough in a day. Acres were traditionally 220 yards, the length an average ox could pull without stopping, by 22 yards, a chain and width needed to turn a team of oxen along the headland and bring them back into work.

aftermath: see *foggage*.

beeves: synonymous with steers – cattle (castrated if males) being reared for meat.

belted: see *colour-sided*.

Blue Grey: the first-cross hybrid between a Galloway cow and a Cumberland White Shorthorn bull.

broken-coated: synonymous with rough-coated, meaning having a coarse waterproof outer coat.

byre: northern word for cow-house.

carucate: the amount of land tillable by a team of eight oxen in a season. Equal to 8 oxgangs and 4 virgates – usually 120 acres. See *hide*.

chain: unit of length of 22 yards, equal to four rods of 5½ yards.

colour-sided: white colouring that makes the animal look as if a sheet has been thrown over its back, particularly in the now extinct Sheeted Somerset and the Dutch Lakenvelder. Similar to belting in the Galloway.

dewlap: the pendulous flap of skin and flesh that hangs below the lower jaw or neck and 'laps the dew'.

dun: a café au lait colour.

etterlin: Scottish word for a year-old heifer in calf.

fat beast: a beast whose body has fully matured so that its bones and muscles have grown to their fullest extent and it has laid down fat either between the tissues or as a covering over the body. See *finishing*.

finching: a characteristic white marking that runs from the belly to the dewlap and on to the head, then along the spine and tail. Particularly noticeable in the Gloucester and Hereford, and also in the animals in the cave paintings at Lascaux.

finishing: the stage at which the animal begins to add fat to the foundation of bone, offal and lean meat or muscle it has previously built up, in that order.

foggage: the fresh green regrowth after a crop of grass has been mown for hay. *Aftermath* is the regrowth after a crop of grass has been taken for silage.

freemartin: the female of a pair of male and female twin calves that is almost invariably barren, and is kept as a cow free for fattening; from the Scottish 'mart' meaning a fattened ox.

furlong: 10 chains or 220 yards, and the length of a furrow – a 'furrow long'. Eight furlongs equal a mile, which is 1,760 yards and 1,000 Roman paces. A Roman pace was two of our paces because they counted the left and the right pace together as one.

genetically modified crop (G.M.): crop grown with seed that has been genetically modified to have a particular characteristic, usually resistance to herbicide.

genotype: the complete heritable genetic identity of an individual; also refers to a particular gene or set of genes that determine one of its characteristics.

haplogroup: a group of similar haplotypes that share a common ancestor.

haplotype: a group of genes in an organism that are inherited together from a single parent.

headland: that part of a field left unploughed at both ends of a furlong where the ox team was turned. The headland was left to the last and ploughed round and round rather than up and down.

heriot: OE *heregeat* ('war gear'). Originally a death duty in medieval England by which a nobleman's accoutrements of warfare – horses, swords, shields – had to be returned to his lord who had provided them. Later, when noblemen provided their own equipment, it was transmuted into a money payment. It then became the right of a superior lord to take his tenant's best beast upon his death in return for his heirs being allowed to remain in occupation of his land. Heriot survived into the twentieth century and sometimes touched the great and the good. Lord Rothschild held a copyhold estate under the Warden and Fellows of New College, Oxford, as his superior lords. They had the right at his death to claim as a heriot his best beast, which, in the case of so distinguished a racing man as Rothschild, might have been worth £20,000 or more. Heriot was abolished in England in 1922. See *carucate*.

hide: 4 virgates or yardlands, about 120 acres, and the amount of land that would support a family. It varied according to the quality of the land.

humus: the accumulated decayed organic matter in soil that gives it fertility.

hybrid vigour or *heterosis*: the effect of the first crossing of two pure breeds, which combines and enhances the best characteristics of each parent and subordinates their less desirable ones in an animal that is superior to both its parents. Every time the cross is made it produces an animal with the same characteristics so long as both parents have been bred pure for long enough. Most enhanced agricultural production around the world depends on hybridization: poultry, eggs, vegetables, fruit and so on.

linseed cake: the pressed residue after the linseed oil has been extracted from flax seed. The cake is used for cattle feed and contains valuable polyunsaturated fatty acids, which have considerable health-giving properties for both the animal and its human consumer.

marbling: the fat that is accumulated within the muscles and tissues of an animal and makes the meat more succulent.

moiled: a Celtic word meaning bare and used to describe polled, i.e. hornless, cattle.

night-stance: a place where drovers rested and fed their cattle overnight.

nook: a quarter of a hide – about 30 acres.

oxgang: the amount of land that could be worked by one man with one ox in one season – usually 15 acres, but it depended on the lightness of the soil.

peck: two gallons (see table of imperial measurements below).

perch: a square rod, i.e. 30¼ square yards.

phenotype: a description of an individual's actual physical characteristics, such as eye and skin colour, also inherited characteristics, such as a propensity to certain diseases.

poaching: describes the effect of grazing animals trampling or plunging land when it is wet and damaging the surface by leaving deep footprints in it.

points: the dark (usually black or red) marking that appears mainly on the nose, ears, feet of some cattle notably the White Park.

polled: hornless. A dominant genetic trait in cattle.

prepotency: the quality that one parent has of impressing its offspring with recognizable and strongly heritable characteristics that seem to override the influence of the other parent.

quarter: one of the four milking compartments of a cow's udder with a teat attached. Also a quarter of a hundredweight (28 lb) and an eightieth of a ton.

quey or *quay*: Scottish word for a heifer or a young cow that has not yet had a calf.

rod: a unit of length equal to 5½ yards. It probably originated from the length of an ox goad, the long stick that the ploughboy carried to urge on the team and with which he measured the width of a piece of land to be ploughed, which was a chain of 22 yards (four rods), the length of a cricket wicket. Forty rods make a furlong.

shambles: the place where animals were slaughtered.

slow feeding: a late-maturing beast that eats more than others to gain the same weight.

sock: white marking on one or more feet that resembles a sock.

stall-feeding: fattening a beast by tying it in a stall and bringing all its food to it.

store cattle: growing cattle that are not ready to be finished.

suckler cow: a cow kept to rear her beef calf by suckling it.

sward: the cover of herbage on a field.

virgate: the amount of land tillable by two oxen in a season. Usually 30 acres, but it varied with the soil, and equal to a quarter of a hide. See *nook*.

British imperial measures

Capacity

1 pint = 4 gills = 20 fluid ounces (0.568 litre)
1 quart = 2 pints (1.136 litre)
1 gallon = 4 quarts = 8 pints (4.546 litre)
1 peck = 2 gallons
1 bushel = 4 pecks = 8 gallons

Weight

1 ounce (oz) = 16 drams
1 pound (lb) = 16 oz (0.454 kg)
1 stone = 14 lb
1 quarter = 28 lb
1 hundredweight (cwt) = 4 quarters = 112 lb
1 ton = 20 cwt = 2,240 lb (1.016 tonne)

Money (pre-February 1971)

penny (d) = 4 farthings (d = Roman denarius)
1 shilling (s.) = 12d
1 crown = 5s.
1 pound (£) = 4 crowns = 20s.
1 guinea = 21s.

Select Bibliography

Whenever I stray, I find myself drawn back to William Youatt, whose great book *Cattle* (1834) is a treasure trove of information. The great man is not always right, but he is always a fascinating guide and chronicler of cattle and where they came from, between the end of the Napoleonic Wars and the accession of Queen Victoria. Robert Trow-Smith's two-volume history of British livestock is invaluable. His writing is a joy to read – a towering achievement, considering how dull his subject could be in the hands of anybody less of a master. I have found myself referring regularly to *Two Hundred Years of British Farm Livestock* by Professor Stephen Hall and Juliet Clutton-Brock, which contains in short chapters a deceptively comprehensive and knowledgeable account of British cattle breeds. Too little has been written on English farming, but for its general history the student could do worse than consult the masterly narrative of *English Farming Past and Present* by Rowland Prothero (later Lord Ernle).

The following selection might be of interest to readers who wish to pursue some of the things this book touches on.

Bahn, Paul G., and Vera B. Mutimer (eds), *Chillingham: Its Cattle, Castle and Church*, Fonthill Media Ltd, Stroud, 2016

Bates, Cadwallader John, *Thomas Bates and the Kirklevington Shorthorns: A Contribution to the History of Pure Durham Cattle*, Robert Redpath, Newcastle upon Tyne, 1897

Berry, Wendell, *The Unsettling of America*, Counterpoint, Berkeley, 1977

Blatt, Harvey, *America's Food*, MIT Press, Cambridge, Massachusetts, 2008

Clutton-Brock, Juliet, *A Natural History of Domestic Mammals*, Cambridge University Press and British Museum (Natural History), London, 1987

Colgrave, Hilda, *St Cuthbert of Durham*, G. Bailes & Sons, Durham, 1947

Collis, John Stewart, *The Worm Forgives the Plough*, Charles Knight & Co., London & Tonbridge, 1973

Dixon, H. H., *Field and Fern (South)* and *Field and Fern (North)*, Vinton & Co. Ltd, London, 1865

Dixon, H. H., *Saddle and Sirloin*, Vinton & Co. Ltd, London, 1870

Dobie, J. Frank, *The Longhorns*, University of Texas Press, Austin, 1994

Egan, Timothy, *The Worst Hard Time*, Houghton Mifflin Company, New York, 2006

Fiske-Harrison, Alexander, *Into the Arena*, Profile Books, London, 2011

Fraser, A., *Research for Plenty*, London, 1953

Graves, Robert, *The White Goddess*, Farrar, Straus and Giroux, New York, 1952

Haldane, A. R. B., *The Drove Roads of Scotland*, David & Charles, Newton Abbot, 1973

Hall, Stephen J. G., and Juliet Clutton-Brock, *Two Hundred Years of British Farm Livestock*, British Museum (Natural History), London, 1989

Hemingway, Ernest, *Death in the Afternoon*, Jonathan Cape, London, 1935

Holderness, B. A., and Michael Turner (eds), *Land, Labour and Agriculture, 1700–1920: Essays for Gordon Mingay*, The Hambleton Press, London, 1991

Harnaday, William T., *The Extermination of the American Bison*, Government Printing Office, Washington DC, 1889

Housman, William, *Cattle Breeds and Management*, Vinton & Co., London, 1915

Howard, Sir Albert, *Farming and Gardening for Health or Disease*, Faber & Faber, London, 1945

Hudson, W. H., *Far Away and Long Ago*, J. M. Dent and Sons, London, 1939

Kerry Cattle Society of Ireland, *Kerry Cattle, A Miscellany*, Killarney, Co. Kerry, 2000

Linklater, Andro, *Owning the Earth, The Transforming History of Land Ownership*, Bloomsbury, London, 2014

Logsdon, Gene, *The Man Who Created Paradise*, Ohio University Press, Athens, 1998

Lord Ernle, *English Farming Past and Present* (6th edn), Heinemann, London, 1961

MacKillop, James, *Myths and Legends of the Celts*, Penguin Books, London, 2005

Porter, Valerie, Lawrence Alderson, Stephen Hall and D. Phillip Sponenberg (eds), *Masons' World Encyclopaedia of Livestock Breeds and Breeding*, CABI Publishing, Wallingford, Oxfordshire, 2016

Quinney, Richard, *Borderland: A Midwest Journal*, University of Wisconsin Press, Madison, 2001

Roebuck, Peter, *Cattle Droving through Cumbria*, Bookcase, Carlisle, 2015

Salatin, Joel, *The Marvelous Pigness of Pigs*, Faith Words, Hachette Book Group, New York, 2016

Salatin, Joel, *The Sheer Ecstasy of Being a Lunatic Farmer*, Polyface Inc., Swoope, Virginia, 2010

Scruton, Roger, *Green Philosophy, How to Think Seriously About the Planet*, Atlantic Books, London, 2011

Stout, Adam, *The Old Gloucester: the Story of a Cattle Breed*, Alan Sutton Publishing Limited, 1980 (revised 1993, the Gloucester Cattle Society)

Street, A. G., *Farming England*, Batsford, London, 1937

Thornton, Clive, *Red Rubies*, Devon Cattle Breeders' Society, Manchester, 1993

Thrower, W. R., *The Dexter Cow*, Faber & Faber Ltd, London, 1954

Trow-Smith, Robert, *A History of British Livestock Husbandry (1700–1900)*, Routledge & Kegan Paul, London, 1959

Trow-Smith, Robert, *A History of British Livestock Husbandry to 1700*, Routledge & Kegan Paul, London, 1957

Ward, Andrew, *No Milk Today: the Vanishing World of the Milkman*, Robinson Books, London, 2016

Williams, William Carlos, *In the American Grain*, MacGibbon & Kee, London, 1966

Wilson, James, *The Evolution of British Cattle*, Vinton & Co., London, 1909

Wolff, Francis, *50 Reasons to Defend the Corrida* (trans. Barbara Ann Sapp), Padilla Libros Editores y Libreros, Sevilla, 2010

Wright, N. C., *Ecology of Domesticated Animals*, London, 1954

Youatt, William, *Cattle: Their Breeds, Management and Diseases*, Robert Baldwin, London, 1834

Young, Arthur, *Travels in France*, George Bell & Sons, London, 1905

Young, Arthur, *A Six Weeks' Tour through the Southern Counties of England and Wales*, W. Nichol, London, 1768

Young, Arthur, *A Six Months' Tour through the North of England*, Strahan, London, 1770

Illustration Credits

'Vaccine-Pock hot from ye cow' by James Gillray, 1802 (*Library of Congress*); Durham Cathedral sculpture (*By kind permission of The Chapter of Durham Cathedral; photograph by Philip Walling*); Holstein heifers (*Pixabay*); Hereford bull (*Jim Guy/Shutterstock.com*); South Devon bull and cow (*Simon Burt/Shutterstock.com*); London butchers' cutting names and prices (*From Robert Trow-Smith's* History of British Livestock Husbandry 1700–1900, *Routledge and Kegan Paul Ltd, 1959*); Galloway cattle (*Wayne Hutchinson/ Alamy Stock Photo*); Kerry cow (*AndreAnita/Shutterstock. com*); Buffalo (*Agricultural Research Service, United States Department of Agriculture*); *The Wild Cattle of Chillingham* by Edwin Landseer, 1867 (*ART Collection/Alamy Stock Photo*). All other photographs taken by the author.

While every effort has been made to contact copyright-holders of illustrations, the author and publisher would be grateful for information about any illustrations where they have been unable to trace them, and would be glad to make amendments in further editions.

Index